COMPENDIUM
OF SCIENCE
DEMONSTRATION-RELATED
RESEARCH
FROM 1918 TO 2008

COMPENDIUM
OF SCIENCE
DEMONSTRATION-RELATED
RESEARCH
FROM 1918 TO 2008

David M. Majerich
and Joseph S. Schmuckler

To order additional copies of this book, contact:
Xlibris Corporation
1-888-795-4274
www.Xlibris.com
Orders@Xlibris.com
46107

CONTENTS

DEDICATION

In general, we dedicate this work to science educators and researchers who wish to positively impact students with whom they interact. In particular, this work is dedicated to science educator and science demonstrator extraordinaire, Dr. Schmuckler, and Chemistry Nobel Laureate, Dr. Alan G. MacDiarmid (1927-2007), who encouraged and mentored this work.

INTRODUCTION

Engaged in selecting and determining the effectiveness of our science teaching methods, we, as science educators, need to continually evaluate our teaching practices in relationship to student learning outcomes. A way to accomplish this is by using the scholarly literature (e.g., anecdotal information and ideas reported by science educators on teaching methods) and existing empirical research (e.g., findings based on cogent research questions, designs, and methodologies) as indicators to gauge our successful teaching. In doing so, we are challenged to continually monitor and revise our teaching methods in light of students' learning outcomes. As less effective methods are replaced with research-proven methods (NRC, 1996, 2006), our repertoires can be transformed from a collection of ideas on how to teach science into an arsenal of strategies when utilized appropriately promote positive student learning outcomes.

However, what do we do if the scholarly literature indicates that a teaching method is beneficial while the empirical research investigating the same method indicates otherwise? We have uncovered this conflict during our investigation of the topic of the demonstration method for science teaching (Majerich, 2004; Majerich & Schmuckler, 2006). While the authors believe that it can be misconstrued that the scholarly literature and empirical research are one in the same, we can show that it is necessary to make a distinction between the two sources of information. Upon review of the scholarly literature and empirical research on the demonstration topics, a salient outcome is that these two bodies of information are only loosely corroborated. From this point on we use the terms literature and research when referring to each of these bodies of information.

In this publication, we share highlights from our journey into the literature and research on the effectiveness of using science demonstrations for the teaching and learning of science (biology, chemistry, physics, earth sciences) predominantly at the high school, college, and university levels. We offer what we believe are two fundamental reasons for the dissonance between the literature and research. In addition, we share some ideas discussed in the psychological literature and research on individuals' perceptions of dynamic events (e.g., demonstrations) to help us negotiate the existing conflict as the authors' own research is discussed (Majerich, 2004; Majerich, Fadigan & Schmuckler, 2008; Majerich & Schmuckler, 2007; 2008). The time and effort expended in reviewing the literature and research bases has informed how we now teach science in our course.

Due in part to the large enrollment and limited availability of science laboratories, our course was mandated from the University's Provost's Office to use a participatory demonstration teaching method. With maximum student participation, the professor involved the students in the course in making observations and discussions of each demonstration. Observations were procured from the demonstrations and elicited discussions, data were collected, and explanations were developed and shared, and these served as the basis for the ensuing lecture discussions. Instructional materials and objects were displayed in the front of the lecture hall; images of small materials and objects were projected on a large lecture hall screen via a projection system.

This course has been in existence for the past thirty years; however, the authors wanted to change the way we taught our chemistry course (holding the curriculum and assessments fixed), making it more student-centered and in alignment with constructivist theory (Bodner, 1986; Mintzes, Wandersee & Novak, 1998; 2000). A course textbook served as a curriculum guide, and assessments were designed to measure students' content mastery and related meaningful learning (Ausubel, 2000). Daily quizzes were used to monitor the students' progress in the course, as well as to inform instruction. Three major assessments were used to monitor how students' learning and understanding of science topics changed over time. Instruction on science topics occurred as students were exposed to science demonstrations (102, in total) and related discussions (Majerich & Schmuckler, 2007; 2008). Modifications to how our course was taught were informed by science standards (NRC, 1996; 2006), general demonstration literature and existing research on the same topic.

LITERATURE-DERIVED CONCEPTUALIZATION OF SCIENCE DEMONSTRATION

The use of demonstrations in classrooms as a teaching aid is uniquely characteristic of the teaching of science (Beall, 1996). Basically, a science demonstration is a public show where a presenter highlights and discusses the prominent feature(s) of an object or an event in front of an audience. The science demonstration is a method of teaching; it is generally performed by a teacher. However, demonstrations can also be performed by a student or group of students, by both the teacher and student(s), or by a guest science teacher/scientist. Instruction may be performed by induction or deduction, and may promote student inquiry through students' own questions and explanations, and/or include teacher questions and explanations (Chiappetta & Koballa, 2002; Trowbridge, Bybee & Powell, 2000).

Demonstrations are not limited to physical objects. For example, a teacher may ask the students to hold an imaginary rope, and then by analogy, proceed to describe concepts related to a standing wave as they move their arms (Ogborn, 1996). When viewed in this broader manner, a demonstration can be conceptualized as the interplay of language, gestures, and objects and/or events—real or imagined—between a presenter and an audience (Majerich, 2004; Ogborn, 1996). Thus, a science demonstration allows for a unique experience, consisting of both concrete and abstract elements, between a presenter and an audience.

NINE (9) FUNCTIONS AND TEN (10) MERITS OF SCIENCE DEMONSTRATIONS

Teaching of science with demonstrations is pervasive in the classroom and widely viewed as beneficial. As derived from the literature (Majerich, 2004), science teachers identified up to nine academic functions of demonstrations in the literature. As reported, the functions of demonstrations are to:

- solve a problem;
- explain and make clear by analysis;
- verify, substantiate, and review; supply an application;
- evaluate student achievement;
- create a problem;

- show methods and techniques;
- display objects and specimens; and
- afford students the opportunity to improve observation skills.

This list has been previously published and contains multiple references (Majerich, 2004; Majerich & Schmuckler, 2007). Most science teachers endorse this list, either in part or in full (Majerich, 2004).

When science educators selected demonstrations to function in any of these nine ways, they reported in the empirical research up to ten merits that positively impacted the teaching and learning of science (Majerich, 2004). The ten merits derived from the use of demonstrations are to:

- act as a motivational device/arouse curiosity/gain students' attention;
- convey the teacher's attitude toward the discipline;
- stimulate the thought processes of students;
- challenge students' knowledge claims, naïve conceptions and/or alternate conceptions;
- help students to focus their attention and/or increase their attention;
- help students negotiate theory and experiment;
- help students see abstract science ideas in concrete examples;
- enhance students' learning of science concepts;
- serve as a substitute for laboratory exercises/experiments that are too costly and/or dangerous to students; and
- develop creativity in students and promote cooperation among students/teacher.

This list has been previously published and contains multiple references (Majerich, 2004; Majerich & Schmuckler, 2007). Interestingly, most science educators also endorse this list, in part or in full (Majerich, 2004). However, the ten merits reported from using science demonstrations remain uncorroborated.

It should be stressed that the above lists of functions and merits offered are comprehensive, and compiled from a nearly exhaustive review of the literature and research, respectively. In addition, these lists are not organized in terms of the importance of the functions or merits of using demonstrations, nor do they reflect any specific frequency of use as identified in the literature and research. However, each function and associated merit(s) is characterized as a unique strategy and outcome(s), respectively. To remain in consonance with the conceptualization of demonstrations advanced earlier in this work, these functions and merits are plausible for any classroom. However, there needs to be a continual interplay of gesture, language, observations of an object and/or event, real or imagined, consisting of concrete and/or abstract elements, between the presenter and the audience (Majerich, 2004).

REVIEW OF SCIENCE DEMONSTRATION-RELATED RESEARCH

In published research, the use of science demonstrations for science teaching claiming positive student outcomes is rather dismal. From our efforts in reviewing this research, we revised an existing *Compendium of Demonstration-related Research from 1918 to 2006* (Majerich & Schmuckler, 2006). In summary, when the lecture is accompanied by a few science demonstrations, the majority of students fail to learn from these science events as typically performed. Furthermore, when a science demonstration was discussed with the students, very few students were able to describe the prominent features and importance of it. In fact, many students could not recall the important information from examples involving common laboratory equipment and fundamental science concepts. Interestingly, many students observed aspects of the science demonstrations that were simply not the intended focus, or simply not there. Since 2001, only a few recent studies examining demonstration-related instructional modifications have shown otherwise (Buchanan, Reynolds, Duersch, Lohr, Coppola, Zusho & Pintrich, 2004; Majerich, 2004; Majerich & Schmuckler, 2006; 2007; 2008; Ophardt, Applebee & Losey, 2005).

PAST IS PROLOGUE: WHAT CAN WE LEARN FROM THE PAST?

Our compendium contains the first study published in 1918 and includes almost all other studies published up to 2008. During this time span, educational theorists like Dewey, Skinner, Piaget, Ausubel, Bruner, Gagne, Novak, Vygotsky, to name a few, have suggested ways for teachers to improve their students' learning of course material in the classroom. However, a common tenet relating teaching and student learning that resides within all of them seems to have been overlooked by many researchers. *Namely, students will fail to learn from an event when exposed to it only once.* For instance, Ausubel

(2000) stated that "learning and retention can be demonstrated only rarely in the absence of frequency, as for example, when the stimulus (learning) material is exceptionally vivid or unusual" (p. 16). He also maintained this position in his earlier work (Ausubel, 1963). Most researchers stated directly or implied that the students viewed demonstrations only once. In addition, there is virtually no indication as to how these events were integrated, if at all, into the students' experiences, in and out of the classroom. The authors believe there are two fundamental problems linked to the students' failure to learn from demonstrations: (a) exposure to a demonstration only once; and (b) lack of integration of experiences. The authors now share with you the result of a collective venture into the literature and research that informs how our student-centered class should be managed and taught. Upon review of this compendium covering ninety-years, we hope that you will find ideas that can be modified and used when using demonstrations in your own classrooms.

David M. Majerich
Joseph S. Schmuckler

COMPENDIUM OF SCIENCE DEMONSTRATION-RELATED RESEARCH FROM 1918 TO 2008

OVERVIEW

This review of research is presented in the form of a compendium of research efforts from 1918 to the present and is comprised of three chapters: (a) *Experimental Studies Comparing the Lecture Demonstration Method and the Individual Laboratory Method of Teaching Science (1918-1989)*; (b) *Further Types of Demonstration-related Comparison Studies (1958-2008)*; and (c) *More Recent Demonstration-related Non-Comparative Studies (1980-2008)*. It was constructed from an intense and nearly exhaustive review of the research. Many research studies contained within this compendium provided fruitful insight into locating additional studies and original references to other studies, that were difficult to ascertain otherwise (Bates, 1978; Cunningham, 1946; Gattis, 1995; Glasson, 1989; Garrett & Roberts, 1982; Kraus, 1997; Stuit & Englehard, 1931; Swafford,1989; Majerich, 2004; Yager, Engen, & Snider, 1969).

The present authors felt that a box-score analysis of these studies was neither a suitable nor possible representation of past research studies. Since 1918, student characteristics (e.g. elementary school, middle school, high school, post-secondary school) have changed considerably, research methods to evaluate the effectiveness of teaching strategies have changed, and conceptual considerations (i.e. instruments for quantifying or assessing desired constructs) have evolved as well. This historical record of research studies could greatly benefit science educators in the future. In addition, references are formatted within the manuscript so that the reader can view a chronology of names with dates and how research efforts involving science demonstrations have evolved over time. This is a work-in-progress and appears here as a revision to an earlier compendium (Majerich, 2004; Majerich & Schmuckler, 2006). All comments, including additional research studies that should be included in this compendium, should be directed to the first author.

CHAPTER 1

EARLY EXPERIMENTAL STUDIES COMPARING THE LECTURE DEMONSTRATION METHOD AND THE INDIVIDUAL LABORATORY METHOD OF TEACHING SCIENCE (1918-1989)

David M. Majerich and Joseph S. Schmuckler

The very first published studies (as cited in Downing, 1931) that tried to prove the effectiveness of demonstration and laboratory methods in science were offered by Wiley (1918), Cunningham (1920, 1924), Phillips (1920), Cooprider (1922, 1923), Kiebler and Woody (1923), Anibel (1923,1924), Carpenter (1925), Walter (1926, 1930), Knox (1927), Nash and Phillips (1927), Pugh (1927), Horton (1928), and Hurd (1929). Downing scrutinized these investigations and found that for these seventeen studies a total of thirty-six schools and forty-six teachers were involved. The students were separated into two groups; each student was matched with a student in the opposing group based on his/her intelligence quotient score or a combination of intelligence quotient score and other student-related academic characteristics. Each group of students in each study received science instruction, either by the lecture demonstration or by the individual laboratory method. It was necessary to insure that the language associated with the lecture demonstration was identical to the language used for the written instructions for each of the laboratory exercises. However, in some of these studies, the discussion method was also utilized for both sections during science instruction. All of the groups of students were tested immediately after receiving science instruction by one of the two methods; in addition, some of the groups of students were tested on the experiment up to two weeks (Nash & Phillips, 1927; Kiebler & Woody, 1923), four weeks (Wiley, 1918), one month (Cunningham, 1920; Cooprider, 1923), three months (Cunningham, 1924), or five months (Anibel, 1924) after receiving instruction by one of the two methods.

Downing reported that "[i]t is evident that as far as the immediate tests go there is a large preponderance of evidence in favor of the demonstration method," (p. 318) pointing out that in only three of the studies, specifically those performed by Wiley (1918), Horton (1928) and Hurd (1929), there was little evidence to support the "superiority of the laboratory method" (p. 318). When students were asked to respond to questions pertaining to the set-up of the apparatus, they performed better on the delayed tests with respect to set-up-related questions if they were actually afforded an opportunity to manipulate the apparatus for themselves. Overall, he noticed that when the tests contained questions that required students to identify the purpose of the experiment, to describe the occurrences in the experiment, and to indicate what the experiment proved, the lecture-demonstration was the better method.

Also in this summary, Downing offered some sobering information with respect to the use of experiments in science education. The results obtained from the Cooprider (1923) study showed that only one-fourth of any group of students had an understanding of what the experiment showed on the immediate tests; on the delayed tests, only one-twelfth of any group of students could identify what the experiment "proved" [sic] (Downing, 1931, p. 319). Similarly, the results from the Walter's (1926, 1930) study showed that, on a delayed test, less than one-fifth of any group could recognize what the experiment "proved" [sic] (Downing, 1931, p. 319). Based on these limited findings, Downing advised that "experimental work, whether done by the laboratory method or by the demonstration method, under good teachers is relatively futile" (p. 319). He continued by adding,

> [a]n experiment is a question asked of nature, and, if four-fifths of the pupils fail to receive any answer, what is the use of asking the question? Experimental work in science may be interesting busy work, but its value as a means of teaching science is evidently not great. Teachers apparently take it for granted that the experiment proves [sic] something to the pupils because it is meaningful to the teacher. That evidently is an error, and much more drill then is customary now must be given to make the experimental work significant. (p. 319)

Adding to his conclusions, he warned that overgeneralizations should not be made even with those studies where "there is unanimity of opinion" (p. 319) and that the lecture-demonstration method was as effective as the laboratory method for the teaching of secondary school science. In fact, he held that a large amount of information could be passed on to the students with a savings of time and money. Finally, he suggested that the demonstration method would be suitable for the college level, "when the science course is cultural rather than vocational" (p. 319).

Shortly after the publication of Downing's critique of the early research and the associated outcomes of those comparative studies, another summary was published which contained the research efforts (as cited by Stuit and Englehart, 1932) of Wiley (1918), Carpenter (1925), Anibal[sic] (1926), Knox (1927), Nash and Phillips (1927), Horton (1928), and Pugh (1929). In this summary, the reported benefits of the lecture demonstration and the individual laboratory methods as they specifically related to the teaching of high school chemistry were scrutinized. As reported by Stuit and Englehart (1932), the conclusions drawn by these seven research studies comparing the lecture demonstration and the individual laboratory methods can be partitioned into three categories. In their respective studies, the individual researchers concluded that the individual laboratory method is superior to the lecture demonstration method (Anibal)[sic], 1926; Horton, 1928; Knox, 1927); that the lecture demonstration method is superior to the individual laboratory method (Anibal)[sic], 1926; Carpenter, 1925; Nash & Phillips, 1927); and/or that with respect to student achievement, there is no difference between the individual laboratory and the lecture demonstration methods (Wiley, 1918; Anibal[sic], 1926; Carpenter, 1925). It is interesting to note that some educators provided conclusions for each of the three aforementioned categories. For example, Anibal[sic] (1926) maintained that the individual laboratory method is superior to the lecture demonstration with respect to student retention of the science content. In addition, he posited that bright students have a greater likelihood of benefiting from viewing the lecture demonstration. Finally, he indicated that the immediate retention rate of course-related science content by students subjected to either of the two methods is equivalent. Based on their critical analysis of these seven studies, Stuit and Englehart (1932) reported that "the relative merits of the lecture demonstration and individual laboratory methods still seems to be unsolved and as complex as ever" (p. 391). They suggested that more rigorous experimentation, better control of confounding variables, and more reliable testing, were required in order to draw conclusions with a high degree of confidence and generalizability.

Stuit and Englehart (1932) prescribed some advice to future science education researchers who were interested in investigating the influence that instruction delivered by the lecture demonstration and the individual laboratory method had on student achievement in chemistry. From these seven studies, they identified seven criteria upon which these and future comparison studies could be discussed: (a) specification of experimental factors; (b) control of pupil factors; (c) control of teacher factors; (d) control of general school factors; (e) duration of experiment (research, in this case, refers to the length of the study); (f) measurement of achievement; and (g) interpretation of the experimental data. First, they noted that the researchers employed the lecture demonstration and individual laboratory methods differently from teachers during science instruction. For instance, they found that both students and teachers were engaged in questioning during the demonstration method. Other studies indicated that the teacher had not posed any questions pertaining to the demonstration. Studies pertaining to the individual laboratory method were also problematic; experiments were sometimes performed on an individual basis, in dyads, or in small groups. Their recommendation was that the instructional procedures associated with the methodologies be explicated "in seeking to determine experimentally the relative influences of the lecture demonstration and the individual laboratory methods on the achievement of chemistry students" (p. 381). Second, they suggested that guidelines for the selection of the participants for each of the research studies be established so that the comparison groups consist of students with similar characteristics. Since academic achievement was to be measured, they offered that "satisfactory equivalence" (p. 381) of the student characteristics of the comparison groups consist of factors such as intelligence, study habits, age, previous science and mathematics achievement, home environment, sex, race, and physical condition. They found that students in the comparison groups differed widely with respect to these factors. In terms of instructional materials, they recommended that the studies utilize identical equipment and textbooks.

In addition, they suggested that the amount of time that a teacher devoted to a demonstration be equivalent to the amount of time expended by students who performed the same experiment. They recommended that the students who participated in the lecture demonstration and those who performed the individual laboratory experiment "should take their work at the same time of the day in order to control the factor of fatigue" (p. 382). This required the use of two teachers who had similar teaching styles. Subject matter and associated topics should be selected so that students in the comparison groups are exposed to the identical information. For instance, provisions should be established so that students receive equal amounts of time in lecture, recitation, and individual study on the science content. Based on these recommendations, Stuit and Englehart (1932) believed that more research was required in the field. At the time of their published summary, they felt that the promulgation of a superior method for the teaching of chemistry was unjustified.

Fourteen years later, the next summary was compiled by Cunningham (1946) and reviewed studies comparing the effectiveness of individual laboratory and lecture demonstration methods over a period of 25 years. In this review (as cited by Cunningham, 1946), the studies previously identified by Stuit and Englehart (1932) were further critiqued. However, Cunningham (1946) also offered an additional 30 research studies and their associated outcomes of comparison studies across multiple grade levels and science disciplines. One study pertained to the teaching of physical science (e.g. elementary physics) at the elementary school level (Mayman, 1912). Of those studies at the high school level, three were in general science (Scott, 1929; Boretz, 1930; Goldstein, 1937), seven were in high school biology (Cunningham, 1920; Cooprider, 1922; Hunter, 1922; Cooprider, 1923; Cunningham, 1924; Johnson, 1928; DeJarnett, 1934); nine were in high school physics (Phillips, 1920; Kiebler & Woody, 1923; Dyer, 1927; Wilkinson, 1928; Shore, 1929; Brasure, 1929; Walter, 1930; Hix, 1933), and fourteen were in high school chemistry (Wiley, 1918; Anibal[sic], 1926; Carpenter, 1926; Pruitt, 1925; Knox, 1926, 1927; Ewing, 1926; Nash & Phillips, 1927; Horton, 1928, 1929,1930; Pugh, 1929; Erickson, 1929; Van Horne, 1929). At the college level, one study was identified in each of chemistry (Payne, 1931), engineering (White, 1943), and biology (Kahn, 1937).

In reference to the students' immediate recall of experiments and results, Cunningham (1946) reported that there were twenty studies that favored the lecture demonstration method; six studies favored the individual laboratory method, and two studies established that there was no difference between the two methods. Twenty-four of these thirty-seven studies addressed the students' delayed recall of experiments and results; ten studies favored the demonstration method, eleven favored the laboratory method, and three indicated that there was no difference in the methods. In terms of stimulating student interest, the majority of students in three studies preferred the lecture demonstration method to the individual laboratory method. In four studies, the majority of students favored the individual laboratory method over the lecture demonstration method. Fifteen studies pointed out that the lecture demonstration consumed one-fifth to one-half of the time associated with the individual laboratory method. Four studies noted that the individual laboratory method afforded students the opportunity to build the skill of manipulating laboratory apparatus, as compared to the lecture demonstration method, which did not. Seventeen of the research studies addressed one or more of the "elements of scientific thinking" (p. 77). These included,

> amount retained in thought work; making proper conclusions to an experiment; application and the interpretation of the results of an experiment; application of principles learned; ability to think in terms of the science subject; ability to follow the steps in scientific procedure; percent of thought questions answered correctly; method of attack on new problems; scientific attitude; ability to observe; learning a scientific principle; greater carry-over ability; ability to distinguish between fact and superstition; and ability to generalize. (p. 77)

Based on the observed development of these skills in students, the results of twelve of these studies favored the use of the demonstration method, four favored the individual laboratory method, and one study suggested that students could think equally as well utilizing either method. However, even though these studies addressed the skills associated with thinking scientifically, no one study evaluated this in a significant enough manner to adequately address this growing concern (Cunningham, 1946). In addition, researchers reported that the lecture demonstration method was not as expensive as the individual laboratory method; however, there is little evidence to support the claims of reduced expense in these research studies (Cunningham, 1946).

Cunningham (1946) noted that in some of the research studies there were variables that should have remained fixed for the duration of the study but did not; this may have affected the results of the study. For instance, he offered that there is evidence that the results of some studies were impacted upon by variables such as, "the complexity of experiments done and apparatus used; length of the period over which an experiment necessarily extended; size of the laboratory apparatus; closeness of view necessary in observing the results of an experiment; sex of pupils; and the time spent upon an experiment by one method as compared to the time spent of the same experiment by the other method" (p. 71). In addition, he found that some of the researchers reported some variables that may have influenced the results of the study; these factors are "age of pupils; year in school; bright or slow pupils; previous science studied; time of day; temperature conditions; previous mental work on the day of experiment; with or without detailed directions; different or same teacher for control and experimental groups; and who performed the demonstrations—the teacher, or one or more of the pupils" (p. 71). These inconsistencies, Cunningham (1946) proffered, may have impacted the results of the research studies.

> Based on his review of literature, Cunningham (1946) offered, [i]n making final consideration as to what we should do in practice as a result of knowing about the experimental work on this problem, it is well to remember, of course, that all generalizations made after studying factual data are, to some extent, guesses since a generalization always covers more cases than have been actually reported. Therefore, it is probable that no absolute decision on this general problem for all cases and for all time can ever be made. (p. 78)

Even though he found the results of the research to be inconclusive with respect to the effectiveness of the lecture demonstration versus the individual laboratory methods, he offered some hypotheses regarding procedures related to the teaching of laboratory-based science. For instance, he suggested,

> [w]hen ordinary written information tests are to be used in the evaluation of the results of teaching and when all other important factors in the teaching situation are, or can be made, favorable, consider the use of the lecture demonstration method if: the learning involved in connection with the exercises is complicated and difficult; the apparatus used is complicated, difficult to manipulate, or expensive; the apparatus used is sufficiently large to be seen at a distance; the pupils are likely to make mistakes, when working alone, in determining and interpreting the results after an exercise has been completed; a large amount of a subject matter must be covered in a limited time. (p. 79)

Similarly, he provided some suggestions for the use of the individual laboratory method for the teaching of science; he concluded,

[w]hen ordinary written information tests are to be used in the evaluation of the teaching results and when all other important factors in the teaching situation are, or can be made, favorable, consider the use of individual laboratory work if: the exercises are short and easy—not complicated as to learning involved or apparatus used; caring for individual differences seems especially desirable; the results can be easily seen and interpreted, by the pupils working alone, after the exercise has been performed. There are some data that indicate that the individual laboratory method may have merit in easy laboratory exercises even though they extend over a long time—especially if several observations must be made over a period of days. A few data were found which indicate that girls made a little better use of the individual laboratory method than boys. (p. 78)

He stressed that the individual laboratory method should be utilized if the teacher's objectives are related to inculcation of laboratory skills, problem-solving ability, or to develop laboratory resourcefulness. In addition, he recommended that both methods be used to provide for the students a plethora of science-related experiences and to increase student interest. Finally, he offered the idea that "both methods can probably be used to advantage in its development" (p. 79). However, he added that much more work was required to determine which procedure would optimize student learning.

In 1961, Bradley (1965) conducted an experimental study in which he compared the lecture demonstration and the individual laboratory methods in association with the teaching of a university physical science course. The participants in this study were first-year and transfer students who were enrolled in the final term of a three-term sequence. Eight groups of students were selected for this study. However, the groups of students were studied at different times during the 1960-1961 academic terms. During the spring term of 1960, one teacher taught one group of students via the lecture demonstration method while another science instructor taught the other group with the same method. Three additional sections of this course taught by this teacher via the individual laboratory method were studied as controls. In addition, during the spring term of 1961, another group of students was taught by another instructor via the lecture demonstration method. Likewise, three additional sections of the course taught by this teacher via the individual laboratory method were studied as controls.

Historically, that course was taught via the individual laboratory method, and class size ranged from 35 to 38 students for each section. The lecture demonstration method represented the experimental portion of the study, and initially, class size ranged from 80 to 82 students per section at the beginning of the term. At the end of the term, class size was 77 for each section. The procedure for teaching this course required that the students participate weekly in a two-hour laboratory, a one-hour discussion, and two one-hour lectures.

The lecture demonstrations were performed by the teachers, in a rigid manner, during the two-hour laboratory period; however, the lecture demonstration lasted for an hour, and the students had the option of returning to the laboratory to revisit the demonstration on their own. For most of the teacher-presented lecture demonstrations, the laboratory equipment used was oversized. This measure was critical to ensure that students in the lecture hall were able to see the equipment and materials. Students who participated in the individual laboratory method performed their own experiments over the course of the two-hour laboratory period. However, for both methods, students were afforded identical reading materials, lecture material, and constant contact hours. The instructors were not required to follow the same pattern of testing. However, they remained consistent with their testing practices for the four groups of students assigned to them for the purposes of this study.

The findings of this study suggested that the lecture demonstration and the individual laboratory methods were equally effective means of teaching, as evidenced by student learning and retention compared to the objectives for the natural science, three-term sequence course. In addition, as part of a plan for further research, Bradley (1965) suggested that there were many methods that were appropriate for the teaching of science; however, it was important to resolve the proper function of each.

At the same time, Strehle (1964) compared the scores on general science achievement tests of students who were taught by the laboratory method with those students who were taught via an enriched lecture demonstration method during a six-week general science program. He noted that the lecture demonstrations were enhanced with programmed instruction, films/filmstrips, overhead transparencies, and models. While there were no significant differences reported between achievement test scores, the author noted that the lecture demonstration was more effective with low achieving students and the laboratory method yielded a wider variation in individual student performance.

The results of the research pertaining to the comparison of the lecture demonstration and individual laboratory methods remained unresolved through the mid-1960's. With respect to the teaching of biology, Coulter (1966) and Sorensen (1966) reported results from their respective studies that were drastically different. Coulter (1966) was interested in determining the effectiveness of science instruction based upon inductive laboratory experiments, inductive demonstration experiments, and deductive laboratory activities. For this particular study, twenty-five ninth grade high school students who were enrolled in a biology course were randomly assigned to one of the three biology classes. The same teacher was responsible for the teaching of all three classes. The author concluded that the inductive laboratory and inductive demonstration methods were as effective as the deductive laboratory activity. Specifically, the inductive methods had not inhibited the students' success in learning facts and principles. Also, he reported the inductive methods were "more conducive to teaching aspects of scientific inquiry, such as cause and effect relationship, making judgments after examining evidence, and evaluation of arguments" (p. 186). The author did note that the students who were afforded the opportunity to perform their own experiments were "more positive in their reactions toward the instruction in their class than were those who watched demonstrations" (p. 186). Overall, no method was determined to be superior to other methods with regard to the teaching of biology for these groups of students.

A few years later Sorensen (1966) reported results from a comprehensive study involving twenty sections of high school biology. Sorensen investigated the relative effectiveness of laboratory-based and lecture demonstration-based patterns of instruction and their impact upon: (a) students' capability to improve their critical thinking; (b) the influence that students' understandings of biology had on their capability to improve their critical thinking; and (c) the influence of students' open—and closed-mindedness on their capability to improve their critical thinking. Students in ten sections received biology instruction that was lecture demonstration-centered. The students in the remaining ten sections received instruction that was laboratory-centered. For all twenty sections of biology, there were sixteen teachers who aided in this study. The author reported that those students who received instruction via the lecture-centered method exhibited greater growth in critical thinking and science understanding. However, the results of this large study and the research design were questioned because of the uncontrolled teacher variable (Yager, 1966). Stuit and Englehart (1932) and Cunningham (1946) informed the science education community of their concern about the teacher variable many years prior to the Sorensen (1966) study. The uncontrolled variable of teacher eroded the research design of this study in particular, as well as many studies prior to this one. Subsequently, the results of such studies were called into question (Yager, 1966).

As a reaction to the Coulter (1966) and Sorensen (1966) studies, Yager, Engen, and Snider (1969) investigated the effect that discussion, discussion-demonstration, and discussion-laboratory approaches had on eight outcomes of instruction. In addition, Yager, Engen, and Snider (1969) emphasized more careful experimental controls, and they explored some additional non-content-related measures previously not pursued in earlier studies. The specific outcomes of interest to the researchers in this study were (a) critical thinking skills; (b) improvement in understanding science; (c) growth in the ability to understand scientists as a group; (d) attitude toward the study of science; (e) changes in knowledge of science; (f) mastery of general biology; (g) general achievement in biology; and (h) ability to use laboratory tools in biology. Sixty eighth grade biology students were randomly assigned to one of the three treatments. Because of scheduling problems, the groups were uneven; however, the groups were equivalent with regard to mean intelligence quotient, science backgrounds, means on pre-test scores, abilities, skills, interests, and attitudes. In addition, the biology course was the same for each group. Identical features for each group were textbooks, time/duration of class periods, examinations, teachers, lesson format, unit sequence, and science content (based on the BSCS Blue Version Course 'Biological Science: Molecules to Man'). The three treatments differed only with respect to laboratory method.

Even though each method was assigned a different teacher, the researchers took great strides to ensure that the teachers selected for the study were matched on experience, education, and dedication. In addition, the teachers participated in each of the treatment groups as they were rotated every four weeks, usually after an examination or at the end of a unit. The matching of the teachers on multiple characteristics and the rotation of the teachers were thought to limit the effect of the teacher variable on the instructional outcomes.

The students assigned to the discussion-laboratory group were the only students who were afforded the opportunity to gain experience by direct manipulation of all of the science-related activities either individually or in groups. Students were given opportunities to discuss their results with others in the class. In the discussion-demonstration group, students were shown one experiment at a time; in this case, the teacher and/or students were the demonstrators. The experiments in this group were identical to those performed by students in the discussion-laboratory group. However, students in the discussion-demonstration group were merely observers of the experiments, and the students as a whole acquired only one set of data from the experiment. Any discrepancies in the experimental data were addressed by the teacher through discussion. In the discussion group, the students merely discussed the results from the experiments. The teacher supplied the experimental information to the students. In this case, the students were not given the opportunity to see or manipulate the apparatus associated with the laboratory exercise, and they did not perform associated laboratory exercises for data acquisition.

However, the students were responsible for interpreting data, drawing conclusions, and relating this information to other situations. Finally, these students discussed experimental design and additional experiments without following through on these investigations. The results of this study suggested that students who participated in the discussion-demonstration or discussion-laboratory groups developed more skills than those students who participated in the discussion-only group. For all other measures, there was no difference among the groups. In response to the Yager, Engen, and Snider research, Shulman and Tamir (1973) offered,

> [h]ave we succeeded in measuring all possible outcomes? Have we not fallen into the trap of using available standardized instruments which have not been designed specifically to test the outcomes of laboratory work and therefore have failed to discriminate and to indicate existing although hitherto unnoticeable differences? (p. 1121)

In summary, these authors challenged the science education community to find better ways to measure and compare subtle differences in students' outcomes after receiving instruction from the demonstration and/or laboratory methods.

The next experimental study to appear in literature was completed by Bybee (1970). In this study, the researcher compared the lecture demonstration method and the individualized laboratory method in a college earth science course. One hundred nine students were assigned to one of the two treatments. Those students who were assigned to the lecture demonstration treatment were required to convene three times a week for one-hour lecture periods that were enhanced with demonstrations and films. Students who received instruction via the individualized laboratory method were required to meet two times per week for one-hour lecture periods; in addition, these students were also required to participate in a two-hour laboratory exercise once a week. The researcher reported that there were no significant differences in comprehensive earth science examination scores between the groups. However, there were significant differences reported on affective measures related to the class for those students who participated in the individualized laboratory exercises.

During the same year that the Bybee studied was published, Peiper and Sutman (1970) offered a brief historical analysis of the use of the demonstration method in the teaching of biology. The overarching question that these authors attempted to answer in this manuscript was: "In the absence of conclusive educational research, what factors must be considered before deciding whether the demonstration or laboratory method will be used?" (p. 83). Although their brief review of research on the use of science demonstrations for the teaching of science produced the same inconclusive results echoed in the science education community previously, the authors offered "certain factors [that] have influenced the trends in the use of demonstrations in teaching biology" (p. 83). They stated that the changing role of demonstrations for the teaching of biology had occurred because of: 1) the nature of biology; 2) the social and economic climate of the time; and 3) the prevailing philosophy of education. In summary, the authors reported "the changing role of the demonstration method has not been as much the result of positive educational research rather perhaps it results from the inconclusiveness of the research" (p. 86).

Over 20 years later, Okpala and Onocha (1988), as reported anecdotally by Bonwell and Eison (1991), conducted a single study that

> clearly show[ed] that students who actively engaged in laboratory experiments designed to illustrate specific principles of physics had less difficulty learning those principles than students who merely saw a similar demonstration illustrating the principle given during a lecture. (p. 12)

However, it was not made clear whether these differences were statistically significant.

Glasson (1989) attempted to determine if the hands-on laboratory method and the teacher demonstration method had any impact on the students' declarative (factual and conceptual) and procedural (problem-solving) knowledge achievement in relation to reasoning ability and prior knowledge. The sample for this study was comprised of 54 ninth grade students enrolled in a physical science course. The students were randomly assigned to one of two treatment classes: a hands-on laboratory method class or a teacher demonstration method class. To control for teacher effect, teachers with similar teaching experience and style were rotated to ensure that they spent an equal amount of time with each of the classes. The physical science three-week unit focused on the topic of simple machines.

Each treatment group experienced the same topics of inclined planes, levers, and pulleys. In the hands-on laboratory classroom, students worked in dyads or triads, were required to construct the apparatus, and collected pertinent data from their manipulations of the laboratory equipment. The procedures for constructing the apparatus and data collection were indicated on laboratory handouts. Students manipulated spring balances and meter sticks to gather data that pertained to the forces of effort and resistance and the distances of effort and resistance, respectively. In addition, these students were given the opportunity to make causal inferences, perform algorithmic calculations such as, input/output, mechanical advantage and friction loss, and answered written questions based on their calculations.

The students in the teacher demonstration classroom were required to make observations of the same laboratory experiments performed in the hands-on laboratory classroom. However, the teacher collected the data and published it on a chalkboard. These students received laboratory handouts that did not include the procedures for apparatus set-up and data collection. Any causal inferences that were generated from students were based on observations of the teacher's manipulation of the equipment and materials. Finally, students were asked to record the data, and they performed similar calculations and answered the identical questions to those administered in the hands-on laboratory classroom.

The findings of this study indicated that the hands-on laboratory method and the demonstration method had no effect on the students' declarative knowledge achievement scores. The students learned the facts and concepts associated with the laboratory experiments and demonstrations equally well, regardless of whether they manipulated the materials. However, Glasson (1989) found that the students' prior knowledge, not their reasoning ability, was significant for the acquisition of concepts. Scores on a computational word-problem test were significantly higher for students in the hands-on classroom as compared to those students in the other group. Last, he discovered that a student's cognitive reasoning ability and prior knowledge could be utilized to predict his/her achievement on a computational word-problem test.

CHAPTER 2

FURTHER TYPES OF DEMONSTRATION-RELATED COMPARISON STUDIES (1958-2008)

David M. Majerich

Frings and Hichar (1958) conducted three laboratory classes in zoology that were enhanced with live animals for demonstrations, lecture, and/or labeled drawings. Although all of the classes addressed the same science content, each was taught by a different instructor. These researchers reported that there were no significant differences in the students' achievement across classes. Although the use of animals aroused the students' interest during the laboratory classes, there was dissension among the students related to the appropriateness of the use of these live animals for this educative purpose.

Another research study compared four methods associated with the teaching of college biology to 924 non-major students. Dearden (1962) reported that all of the students received identical instruction during the lecture portion of the course. Students were divided among one of four treatments. Some students performed individual laboratories, some students witnessed demonstration laboratories, some students performed workbook exercises, and some students were required to write term papers. With respect to scores related to biological content knowledge, scientific thinking, or biological attitudes, there were no significant differences reported for any of the methods of instruction.

In 1975, Oliver investigated the efficiency of three methods of teaching high school biology; these three methods were lecture-discussion, a combination of lecture-discussion and demonstration, and a combination of laboratory exercises and demonstrations. Two classes were assigned to one of the three treatments. While the researcher indicated that the effect of the lecture-discussion was noticeable during the first semester in terms of biology content acquisition, the effect dissipated by the end of the following semester. Overall, there were no significant differences on a researcher-produced comprehensive test in which students' ability to apply biological principles was measured.

In response to the continued trend to return to teacher-centered instruction, Beasley (1982) investigated teacher demonstrations and their effect on student task involvement. The sample for this study consisted of three grade-level eight classes, twelve grade-level nine classes, and nine grade-level ten classes. Each science class ranged in size from 20 to 36 students, with 27 students as the median. Twenty-four teachers volunteered for this study; each teacher was observed executing four lessons, which were approximately one month apart. Of those teachers that participated, nine teachers performed demonstrations. The remaining methods of teaching analyzed were teacher exposition and teacher exposition with a prop. It is important to note that the teacher demonstration, teacher exposition, and teacher exposition with props represented only a segment of the overall lesson plan.

Of a possible total of ninety-six lessons, ninety-one were documented on sound-sensitive videotapes. Six students from each of the classrooms were continuously observed for the duration of the lesson, and trained professionals coded their behavior. A coding system that represented student task involvement was created to indicate that the student: (a) definitely was in the assigned work, task, or expected behavior; (b) probably was in the assigned work, task, or expected behavior; (c) was waiting after completion of the assigned task; or (d) was out of the assigned work, task, or expected behavior. The findings suggested that, for the nine teachers who performed teacher demonstrations, students had significantly higher student task involvement scores than they did during teacher exposition segments with the same students ($p<0.05$). However, student task involvement was not higher when teacher exposition was enhanced with a prop. Beasley (1982) concluded with that fact that "[e]xpository methods were much more common, but in the eyes of the pupils, teachers perform best when they are conducting the classroom demonstration" (p. 790).

In 1983, Eniaiyeju compared the effectiveness of the teacher demonstration method and self-paced instructional method on the concept acquisition and problem-solving skills of college chemistry students. In this study, sixty students who possessed similar characteristics were randomly assigned to two groups. Most of the students had little or no prior involvement with science courses. Thirty students were assigned to the self-paced group, and a teacher acted as a facilitator for them. This group received an instructional package that consisted of three units of chemistry, namely, acids/bases, salts/solubility, and electro-chemistry. Included in this self-paced package were worksheets that described the procedures for performing the unit laboratory activities. For each of the activities, the students were apprised of the necessary materials, statement of purpose and procedures, and a self-check to insure that the performance objectives for the associated activity had been met. Individual students worked at their own pace; however, the teacher provided necessary encouragement and feedback for those students who lagged behind so that they would work more quickly.

Students assigned to the teacher demonstration group also received the self-paced instructional package, but a teacher presented the material to them as well. The teacher wrote the performance objectives for the unit activity on the board, and the students were required to maintain their own records based on the teacher's lectures and chalkboard notes. The teacher also maintained a similar record of the lecture and chalkboard notes.

At the end of the second unit, student achievement for each student in the study was measured from a score on a 54-item objective test. Prior to the third unit, those students who were previously assigned to the self-paced group were reassigned to a teacher-demonstration group; similarly, students in the teacher demonstration group were reassigned to a self-paced group. After completion of the third unit, student achievement was measured from a score on a 20-item objective test. This test was not analyzed because of the possibility of an interaction of methods as a confounding interference. In addition, each student's preference related to the instructional mode was measured from a score on a 16-item Likert scale instrument. Findings of this study indicated that concept acquisition achievement ($p < 0.01$) and problem-solving ($p < 0.001$) achievement were significantly different for the students who participated in the self-paced instructional group when compared to those students who were in the teacher demonstration group. Finally, eighty-one percent of the students favored the self-paced method to the teacher demonstration method.

Several years after Eniaiyeju (1983) published his findings, Swafford (1989) investigated the combined effects of demonstrations and science texts on conceptual change of secondary school students. A new area of study was considered during this study; differences associated with the effects of treatments for males and females were also explored for the first time. The participants in this study were students enrolled in college preparatory science courses from ninth, tenth, and eleventh grades. Eighty-three students participated in the first study; seventy-nine students participated in the second study.

In the first study, students were randomly assigned to one of four treatment groups, namely, Text Treatment, Demonstration Treatment, Demonstration Plus Text Treatment, and Text Plus Demonstration Treatment (italics in original). Students in the Text Treatment group were asked to read about free fall in a text book while engaging the material in the manner that they would if they were studying for a test. The teacher informed the students in the Demonstration Treatment that they would also investigate the concept of free fall. They were supplied with an assortment of materials and measuring instruments for their activity. Prompted by the researcher, the students made predictions and subsequently tested their predictions by performing the demonstration with materials. In the Text Plus Demonstration Treatment, students were asked to read a passage about free fall from their textbooks after which they would watch a series of demonstrations with materials placed before them on a table. All students were given an application problem and a multiple-choice test. Each of these was administered three times—as a pre-test, an immediate post-test, and a delayed post-test. For students who were in the Demonstration Plus Text Treatment, the order of activities as indicated in the Text Plus Demonstration Treatment was reversed.

In the second study, the researcher sought to determine if the results of the first study were generalizable to another concept in science. This time, the concept that was utilized was related to the seasons. Again, students were randomly assigned to one of the four treatment groups previously described; however, the science concept and the materials for the activity were different. Students were provided with similar instructions per treatment group as indicated above.

Swafford (1989) found that in study one, the students' achievement scores were the same regardless of the type of treatment. The results of the change in score analysis indicated that females retained more information on the delayed post-test as compared to the immediate multiple-choice post-test, and males lost information during that time. With respect to the second study, immediate multiple-choice post-test scores for males depended upon their prior knowledge before the treatment. Excluding the Text Plus Demonstration Treatment, there were positive relations between existing knowledge and immediate multiple-choice post-test scores for all treatments. On the delayed multiple-choice post-test scores, males scored the same regardless of the treatment. Higher immediate post-test scores were achieved by females in the Text Treatment and the Text Plus Demonstration Treatment when compared to the immediate post-test scores of females in the Demonstration Treatment. No effect for the treatments was observed for females' scores on the delayed multiple-choice tests. The qualitative data obtained from the immediate and delayed application problem indicated that the students shifted their naïve theories toward becoming more like the scientific theory they were studying. Another study by Marshall (1989) suggests similar findings as those reported by Swafford (1989).

A study was performed by Johnson (1991) that compared the effectiveness and efficiency of videotaped demonstrations, live teacher demonstrations and student-performed hands-on demonstrations for non-science majors. Overall, the purpose of this study was to ascertain if there was a significant difference in student achievement and student attitude toward the methods. In addition, the researcher attempted to determine which method was more cost efficient based on student learning outcomes. The participants in this study were sixty-four students, predominantly non-science majors, enrolled in a semester-long food science and nutrition course. The students were randomly assigned to one of the three treatments. The laboratory component of this course was comprised of three four-week intervals, namely, dairy, meats, and eggs and poultry. Each of the students was exposed to all three treatments for the duration of each of the four-week intervals, but the designation of the sequence of methods for each of the laboratory sections was randomly determined.

In this study, all laboratory students received the following: (a) an introductory discussion; (b) the laboratory exercise; (c) a summary discussion; (d) time for asking questions and review; and (e) a quiz. For those students who were involved with the videotaped demonstration method, they were afforded an introductory discussion, summary discussion, and the actual demonstration performed by the teacher, all of which appeared on the video recording. In some instances, videotaped teacher and student interactions, as well as student-performed hands-on group activities, were shown. For those students who experienced the live demonstration method, it was the teacher who performed the demonstration, giving way to further student-teacher interactions. For those in the student-performed hands-on demonstration group, the students performed the demonstrations themselves with a teacher and teaching assistant available for discussion.

The researcher reports that there were no significant differences in student learning outcomes when the three methods were compared. A student survey was administered prior to the research study, and then again after the semester ended. Results indicated that the students felt that neither the videotaped demonstration method nor the student-performed hands-on method was as efficient and effective as the live teacher demonstration method. Also, based upon the survey, personal interaction was also an important variable in terms of student satisfaction with a teaching method. Finally, in terms of cost and time expenditure for each of the demonstrations that were shown, the videotaped demonstrations were more economical in terms of time and expense compared to the live teacher demonstrations and the student-performed hands-on demonstrations.

A few years later, a multi-method study by Gattis (1995) was performed to ascertain the effect that classroom demonstrations had on student acquisition and retrieval of physics concepts. Specifically, the study was designed to analyze student learning of physics concepts based on different levels of demonstration usage. The participants in this study were students who were enrolled in a calculus-based university physics course. These students were assigned to one of three different treatment methods, which were identified as enhanced usage, typical usage, and no usage (italics in original). As defined by the researcher, the enhanced usage treatment demanded a more active role on the part of the students by requiring them to make predictions and/or explain phenomena to each other. The enhanced demonstrations associated with this method were time consuming; the students were exposed to forty-five demonstrations over the course of the semester; that is, one demonstration per session. Students who were assigned to the typical usage treatment group regularly observed brief demonstrations during the course instruction. Typical demonstrations were presented and carefully explained to the students by the instructor; no provisions were made to encourage the students to verbalize their predictions or understandings of the demonstrations. Students in this treatment group observed sixty-seven demonstrations over the course of the semester; that is, 1.5 demonstrations per session. Students assigned to the no usage treatment group were not afforded the opportunity to observe demonstrations. This group, the control group for the study, was instructed on similar physics content to that which the other groups received; however, instead of demonstrations, the teacher provided example problems related to the physics content that was discussed during the lecture.

For this study, the primary means of data collection was a series of three topic-related quizzes that were administered to each of the students prior to and upon completion of a class on three separate occasions. Each of the quizzes was composed of two questions. The format of the quizzes was multiple-choice. Specifically, each quiz contained two questions, and each question had two sets of concept-related multiple-choice questions. In the first part of the question, students provided their prediction for the outcome of a physics-related phenomenon. Students detailed their explanations of physics-related phenomena in the second part of each question.

Gattis utilized these pre-and post-quiz scores to measure the students' conceptual change with respect to the topics of force and motion, conservation of mechanical energy, and conservation of angular momentum. With respect to the change in scores associated with the force and motion quizzes, students who were assigned to either of the demonstration groups performed significantly better than those students in the no usage group. On the topic of conservation of energy, there was no significant difference observed in student quiz scores. On the topic of angular momentum, it was observed that the students in the enhanced usage and typical usage treatment groups had higher predictions gains, whereas the students in the no usage treatment group showed higher explanation gains.

In addition to the pre—and post-quizzes utilized to measure conceptual change in students, Gattis also relied on interviews to secure qualitative information from the students. Each interview took place within two weeks of the student's observation of the physics-related demonstrations. The purpose of the interview was to determine to what extent students remembered demonstrations, to further analyze students' explanations of the demonstrated natural phenomena, to further determine the role of the demonstration in concept acquisition, and to investigate students' recall of demonstrations and associated concepts. Students in the enhanced usage and typical usage treatment groups who were observed to have conceptual gains on the post-quiz provided the researcher with more complete responses when asked to solve a conceptual problem. However, the language that these students utilized was similar to that of students who were not observed to have conceptual gains on the post-quiz. Overall, there were no differences in interview responses to a conceptual problem supplied to the students in any of the three treatment groups.

Based on these sources of data, Gattis suggested several conclusions regarding the role that demonstrations play for conceptual change; he offered that demonstrations:

> (a) may help to confer belief on a concept that a student finds counterintuitive; (b) may provide visual images that are important components of rich and detailed concepts; (c) may help to explain concepts that have key spatial and temporal relationships; and (d) may provide especially vivid physical examples that are useful in making analogies to other examples and generalizing to a more abstract concept. (p. i)

The next notable study was conducted by Kraus (1997), who focused her research on the effects of lecture demonstrations, as typically practiced, on students' understandings of physical concepts. She also undertook a broader examination of student learning of physics concepts using other modes of instruction (guided inquiry, interactive lecture, small-group tutorials) and their effects upon student achievement during the teaching of a large-enrollment university physics course. In this study, students' understanding of physical concepts was compared among groups of students who experienced demonstrations and groups who were not exposed to science phenomena in this way. The results of the study revealed that students who experienced demonstrations could describe the events better than students who did not experience the demonstration; however, both groups appeared to have the same level of understanding of the physical concepts exhibited by the demonstration. Also, Kraus reported that several students who experienced a demonstration in class retained a memory of the outcome of the demonstrations that did not occur. The evidence presented in this study suggested that many lecture demonstrations, as typically practiced, "do not assist students in their development of functional understanding of the concepts illustrated in the demonstrations" (p. 282). Furthermore, "while it was possible to help students deepen their understanding of the demonstration, the class time might have been better spent addressing some of the underlying difficulties that are not explicitly addressed in [the] demonstrations" (p. 282).

Kraus offered several recommendations for using demonstrations for the teaching of physics in the lecture hall. Demonstrations can: (a) provide a sense for the instructor as to what the students do and do not understand; (b) be used to elicit what students know about science topics; (c) be most effective at eliciting student difficulties with demonstrations and related topics; (d) be effective if "embedded in a carefully planned instructional sequence" (p. 286); and (e) beneficial to students if they precede and are followed by structured questions that guide the students in correctly interpreting their observations of science phenomena. In summary, Kraus stated

> In many of our efforts to improve student understanding of important concepts, we have been able to create an environment in which students are mentally engaged during the lecture. While we have found this to be a necessary condition for an instructional intervention to be successful, it has not proved sufficient. (p. 286)

In addition to creating an interactive learning environment, Kraus advised that the nature of the questions and events that students are asked to think about and discuss in class be given more consideration.

Simultaneous to Kraus (1997), Bowen and Phelps (1997) described a demonstration-based cooperative testing strategy for use in a general college chemistry course. In this study, the authors described three approaches to demonstration-based activities and group testing and their specific use in five different university courses of varying subject matter, student populations, and course support. The three types of demonstration-based assessments are:

(1) As a separate quiz in which students see a demonstration and work together in small groups before handing in their own sheet answering questions about the demonstration after 30 minutes;

(2) As part of an exam in which the demonstration is shown at the beginning of the test and students answer questions about the demonstration that are recorded along with the other test items; and

(3) As a set of questions on an exam that is done 48 hours before the written exam. Students see the demonstration and can work in groups outside of class before turning in the questions as part of the exam. (p. 716)

In this study, the researchers selected the third demonstration-based assessment to evaluate student performance on course examinations.

In brief, the researchers described the format on the day that the demonstration was to be performed. This format consisted of: (a) announcements for the day; (b) presentation of new material; (c) distribution of question; (d) demonstration and recording data; and (e) group (self-selected) work to answer questions. Furthermore, the authors offered a general demonstration-based testing approach that they used during a typical 50-minute class period. Then, students were provided with demonstration-related exam questions 48-hours before the in-class written examinations. Students were afforded opportunities to work with other students outside of class when answering these questions, but each student had to submit his or her own work when due. Within the manuscript, the authors provided five examples of demonstration-based assessments utilized to measure student learning of related terms and content.

Participants in this study consisted of two sections of the first semester of a year-long chemistry course at one of the universities identified within the authors' manuscript. Each section was taught by a different faculty member who had more than five years of teaching experience. These two classes met at the same times during the day, utilized the same text and had the same laboratory requirements. Overall, the students completed five 50-minute examinations. One group, who received the treatment, experienced demonstrations and completed the demonstration-based questions as described above. The results of the study revealed "there was no detectable difference between treatment and control students for content not covered by demonstration-based assessment tasks during the term" (p. 718). However, "treatment students outperformed control students on items for which they had been tested during the term" (p. 718). In summary, the authors reported that the demonstration-based intervention could have a positive impact on a student's future performance on standardized tests that emphasize conceptual understanding. A course survey was also administered to the students to ascertain their perceptions of the assessment formats and study time. Overall, the results indicated that there were no differences between the control and intervention groups on their perceptions of the accuracy of several assessment formats. However, a significant difference was detected on students' perceptions of how much time they spent studying with other students. The treatment group spent twice as much time studying with students outside of class compared to the control group.

The researchers also collected qualitative data from a second university where students were asked, "What changes in the class would you like to see that would help you learn more effectively?" With respect to the demonstration-based quizzes, the students responses were categorized into one of four categories: (a) More frequent quizzing (of a regular quiz); (b) More time (or partial take-home) for quizzes; (c) Unfair that students can get help;

and (d) Avoid linked questions. The researchers reported that these comments led to changes in the approach of using demonstration-based assessment. The greatest changes were "doing the demonstration and giving students several days to work on the questions, and allowing students to choose not to work in groups if they so desire" (p. 719). However, the researchers added, that "based upon observation during class, more than 80% chose to work with other students during class time" (p. 719). The researchers also included some other considerations when using demonstrations and approaches to using demonstration-based assessments in the classroom.

In 2000, Deese, Ramsey, Walcyzk, and Eddy conducted a study in which they examined the effectiveness of demonstration assessments and the impact that they had upon students' critical thinking and deeper conceptual understanding of important chemical concepts. In this study, two intact sections of freshman engineering majors enrolled in a chemistry course provided a control group and an experimental group. All students experienced the same demonstrations. For both groups, the instruction occurred predominantly using traditional lecture accompanied with lecture demonstrations, recitation, and small-group activities. Two instructors taught the sections of the course; each instructor was responsible for only one of the sections taught. The groups were assessed using the same strategies (three non-cumulative 100-point exams designed by the two instructors); however, the experimental group was asked to complete demonstration assessments (which replaced the biweekly course quizzes) every two weeks during the 10-week term. Overall, five demonstration assessments were administered to the experimental group. The procedure for the demonstration assessment was: (a) general rubric was presented to the students; (b) demonstration was presented; (c) students recorded their observations and explanations of what was observed; and (d) student work was collected and a task-specific rubric was used to facilitate classroom discussions. In order to determine the effectiveness of the intervention, the researchers administered the Chemistry Conceptual Assessment to both groups at the end of the course. In addition, both groups were also administered the Scientific Attitude Inventory II as a pre-test and post-test to ascertain their attitudes about science and the pursuit of careers in science. Results revealed that the scores on course exams were higher for the experimental group than for the control group, but these differences were not statistically significant. Results of the attitude inventory showed that all students in the study had equally stable and positive attitudes toward science for the duration of the study. Overall, the researchers concluded that the enhanced conceptual understanding of the experimental group was due in part to the demonstrations assessments. Namely, the demonstration assessments: (a) helped the students to focus their attention on course content to be learned; (b) encouraged a deep elaboration of demonstration-related concepts in the students' own words; and (c) gave students a metacognitive awareness of the thinking process. While this research focused upon student outcomes only, the researchers recommended that future research focus on process measures to verify their speculations.

In 2001, Buncick, Betts, and Horgan compared engagement and academic achievement in an introductory physics course using both demonstrations as typically practiced and a modified demonstration technique; students were exposed to both treatments. In doing so, the researchers focused their attention on improving the engagement of students in the physics course, including underrepresented women and African Americans. The participants were students from both semesters of a year-long physics courses taught first traditionally, and then with a modified method one year later. Here, the professor of the course taught the typical demonstration. "The instructor presents the demonstration and focuses on particular aspects of the event to highlight a physical concept and asks the students to understand only that particular event" (p. 1241). The modified demonstrations, referred to by the researchers as "Road Map Demonstrations" (p. 1241), "are designed to illustrate the interrelationship among concepts from several chapters at one time. Each demonstration looks both forward and backward in course time so that it integrates concepts and establishes continuity from chapter to chapter" (p. 1241). Furthermore, "[w]ith the exception of the first demonstration in the series, each demonstration also ties the concepts that are coming up in the new units to concepts that have been covered in previous units of the course'" (p. 1242). The Road Map Demonstrations were performed with student assistance and used "a series of standard demonstrations as examples of activities that can be used to introduce multiple concepts and tie different sections of the introductory physics together" (p. 1239). Furthermore, the demonstrations served as the context through which the science concepts throughout the course were discussed.

To increase the students' engagement with the demonstrations, they were asked to predict parts of the demonstration outcome and to interpret outcomes based on what they already knew. The results showed that the students in the modified class were more likely to ask questions when invited by the professor (11.5% in the modified course versus 0% in the traditional course) and were more active in initiating their own questions (13.9% versus 6.8%). Also, the students in the modified course responded to 41.8% of the professors questions compared to students in the traditional course who responded to only 23.1%. In terms of student achievement, the researchers reported that students received slightly higher scores on the course tests and final examinations when compared to those in the traditionally taught course. Furthermore, the researchers concluded that the "road map demonstrations can play a role in efforts to 'warm the climate' and encourage retention of women and students of colour in science and engineering majors" (p. 1253).

Another study by Thompson and Soyibo (2000) investigated whether the use of the combination of lecture, teacher demonstrations, class discussion and student practical work in small groups significantly improved students' attitudes about chemistry and understanding of electrolysis when compared to students who did not participate in the practical work experience. In addition, the researchers also investigated whether students' performance on an electrolysis exam was linked to the treatment, gender and post-test attitudes toward chemistry. The participants in this study were two intact groups of tenth grade students enrolled in a chemistry course at one of two high schools. Students were administered an Understanding of Electrolysis test developed by the authors. To ascertain the students' attitudes towards chemistry, the researchers administered the Attitudes to Chemistry Questionnaire. The test and the questionnaire were used as a pre-test and post-test. Overall, nine lessons were taught over a four-week period of time during the course of the study. During this time, the experimental group experienced practical work six times and the control group did not. The study showed that the students who received the combination of lecture, teacher demonstrations, class discussion and student practical work had higher post-test attitude and understanding of electrolysis scores compared to the students who received lecture, teacher demonstration and class discussion. In addition, these differences were statistically significant. The researchers noted that there were no gender differences with respect to the students' post-test scores on the electrolysis test. Although the researchers reported that there was a positive, statistically significant but weak, relationship between the experimental subjects' treatment and post-test electrolysis scores, they suggested that other factors should be examined to account for the difference in scores. They recommended that students' intellectual abilities, cognitive styles, subject preferences, socioeconomic backgrounds, and teachers' qualifications/experience be identified and investigated.

In research that investigated the assessment and enhancement of introductory physics and biology courses, Fagen (2003) focused his attention on the use of peer instruction, classroom demonstrations and genetics vocabulary all aimed at improving student engagement in the lecture and the subsequent learning of course material. The first component of his research contains the results of an international survey on the implementation of Peer Instruction, a learning pedagogy that employs a questioning strategy aimed at improving the student's critical thinking. His conclusions highlight the adaptability and success of the peer instruction framework when implemented in diverse settings. He included recommendations for teachers and researchers who may want to adopt the peer instruction model into their classrooms. The third component of his research was aimed at creating a biology concept inventory aimed for use in determining students' conceptual understanding of basic introductory biology terms and concepts. His research shows that introductory biology students begin their university biology courses with significant gaps in their understanding of terms and concepts, some of which are maintained tenaciously throughout the course. The second component of his research pertains to the enhancement of a university physics course and lecture with the use of physics demonstrations.

The focus of this research was to confirm the results of an earlier study by Kraus (1997), also reviewed in this compendium. The overarching research questions for this component of his study were, "Does student learning from demonstrations depend on the pedagogy with which the demonstration is presented? In particular, does engaging students in the demonstration by asking for them to make predictions about the expected outcome beforehand enhance learning over a traditionally passive presentation? Could learning be further enhanced by asking students to discuss the demonstration in addition to making predictions?" (p. 55).

The participants in this study were students enrolled in a year-long physics course in which the topics of mechanics and electricity and magnetism were discussed. This large enrollment lecture course had 125 and 100 students enrolled in the course for the fall and spring semesters, respectively. Students participated in two 90-minute lectures twice per week taught with the Peer Instruction strategy. In addition, the students also were encouraged to participate in a weekly workshop section comprised of 20-25 students per section. Participation in workshops afforded the students opportunities to develop physics concepts and reasoning skills, supplementing the lecture and text discussions. Furthermore, these workshops were aimed at inculcating problem solving skills within the students.

Overall, there were four workshops for students in which they were exposed to physics demonstrations via one of four demonstration-related strategies. Students were assigned to one of four groups: no demonstration (used as a control where no demonstration was shown), observe (traditional demonstration performed and explained by the instructor), predict (students first predicted the outcome of a demonstration, followed by a traditional demonstration performed and explained by the instructor), and discuss (students recorded their predictions, observed the demonstration performed by the instructor, discussed the demonstration with peers in the class, then the demonstration was discussed by the instructor. A worksheet aided the students in recording their observations and reminded them of the mode when outside of class). Fagen (2003) reported that the "modes varied from week to week with the modes rotating through the sections" (p. 59). This was done in an attempt to account for differences in students' ability across sections. The students experienced demonstrations presented with the four different modes equal numbers of times. At the end of the semester, student learning from the seven mechanics demonstrations and nine electricity and magnetism demonstrations was assessed with a free-response test. The conclusions from Fagen's study were: "increasing the degree of interactivity with which demonstrations are presented appears to also increase student learning from those demonstrations, especially understanding of underlying concepts" (p. 93). In addition, he added, "although different demonstrations and topics show different individual patterns, the overall message that increased engagement—through prediction and discussion—leads to increased learning seems robust" (p. 93). Increasing the students' engagement with demonstrations using the prediction strategy prior to experiencing the related phenomena resulted in an increase in the students' ability to recall and explain that phenomena.

Crouch, Fagen, Callan, and Mazur (2004) repeated a similar study using the four modes (no demonstration, observe, predict, discuss) with a class taught with the Peer Instruction Framework previously described by Fagen (2003). The work reported in this compendium is an extension of the preliminary discussion of results reported by Callan, Crouch, and Mazur (2000) not contained in this compendium. The participants in this study were 133 students enrolled in an introductory physics course designed for premedical students. This class met for 2.5 hours per week, and these students were also encouraged to participate in a weekly workshop. During the workshops, the students were exposed to demonstrations with one of three modes (observe, predict or discuss), or they did not see a demonstration. As was done in the Fagen (2003) study, the mode of demonstration was rotated weekly; in doing so, each student participated in the modes or in the control group the same number of times. A total of seven demonstrations were utilized in this study. At the end of the semester, a free-response test to assess the students' learning from the physics demonstrations was administered. The conclusions stated earlier by Kraus (1997) and Fagen (2003) were echoed by the researchers in this study.

That same year, Buchanan, Reynolds, Duersch, Lohr, Coppola, Zusho, and Pintrich (2004) examined the effects that instructional modifications made to a traditionally taught university chemistry course without a laboratory component had upon students' attitudes or course components and the learning of course content. The original course was taught using the traditional lecture format; the modified course was more student-centered and incorporated a combination of multimedia presentations, interactive questions and demonstrations. The presentations were of the PowerPoint format and afforded the instructor a greater access to visual representations (graphics, computer animations, short video presentations). The interactive questions were of the multiple-choice format. The questions were incorporated into the lecture and students were afforded time to work on the problems. After the students completed the problem, they were asked to vote for their answer by raising their hands. This was done to assess the students' mastery of course content. The demonstrations were incorporated into the lecture to: (a) introduce new course material; (b) help students visualize the chemical concepts; (c) help students in making predictions; and (d) reinforce chemistry applications. During the demonstration presentations, the students were asked to watch the demonstration, think about the phenomena observed, derive an explanation/conclusion about what they saw, and relate the demonstration to the topic being discussed. Students were asked to share their responses to these activities, and the presenter summarized the class consensus for each activity.

The traditional course utilized the same instructional format, content, and weekly quiz which were determined by the graduate student instructor. The discussion format for the modified course was based on a collaborative model where course worksheets were used in an attempt to improve students' problems solving skills. The completion of the worksheets would be done in class; the instructor would facilitate small-group discussions. During these discussions, the students shared their responses with others in the class, and the instructor could highlight important topics and clarify confusing issues or build on students' conceptions. The worksheet was designed to help students better understand the concepts discussed in class. To further emphasize how chemistry impacted the students' lives, the students were assigned four experiments to be completed outside of class.

To ascertain the students' motivations for learning chemistry, self-efficacy and attitudes, all students were asked to complete three surveys. Results revealed no significant difference between students in the traditional and modified courses on motivations for learning chemistry and self-efficacy. Students in the modified course believed the course components were effective strategies, but they did not feel the same way about the experiments. In addition, the students in the modified course felt that all of the instructional modifications were effective in increasing their awareness of chemistry in their lives. The researchers conducted interviews with a sample of students from both courses. During the interviews, students were asked to: (a) reflect on prior science courses and the modified course experience; (b) recall demonstrations discussed in class; and (c) describe chemical concepts underlying demonstrations and real-world applications of the related concepts. A rubric was used to score students' interview responses when asked to identify and recall terms/definitions from course material. Students' understanding of concepts was defined as student ability to explain, integrate, or make connections among the concepts and examples given in the interview. The results from the interview showed that students from the modified course had greater conceptual mastery and deeper understanding of concepts when compared to students from the traditional course.

Finally, in a small preliminary study, Manaf and Subramaniam (2004) investigated the effectiveness of chemistry demonstrations on conceptual understanding and cooperative learning with secondary students at an all-boys school in Singapore. In a single 45-minute session, one class of 25 students was taught the concept of electrochemistry utilizing nine different demonstrations. Another class of 25 was taught the same concept through traditional methods. The authors then compared the results of a post-test, content-based multiple choice test and a 16-question survey designed to elicit students' perceptions of educational effectiveness, the learning environment, and the nature of the demonstrations. The results revealed that the mean score on the concept-based test was higher for the students in the demonstration class. Based upon the student survey results, the authors also note that the demonstration session stimulated interest in learning more about the topic, and promoted thinking skills and an improved cooperative learning environment.

CHAPTER 3

MORE RECENT DEMONSTRATION-RELATED NON-COMPARITIVE STUDIES (1980-2008)

David M. Majerich

A number of more recent demonstration-related, non-comparative studies were conducted between 1980 and 2006. In 1980, Champagne, Klopfer and Anderson utilized a demonstration-observation-explanation (DOE) task when they asked middle school students to make observations and formulate explanations related to a science demonstration where two different cubes, identical in geometry, but different in composition (e.g., plastic, metal), were simultaneously released from the same height above the ground. More specifically, the DOE task required that the students: (a) were presented with the demonstration materials followed by a description of them; (b) were asked to predict the outcome of the science demonstration, as well as articulate the information that informed their predicted outcomes; and (c) were asked to observe the science demonstration, and describe/discuss any differences that exist between their predictions and the outcome of the science demonstration. Overall, these researchers found that the students' observations that were reported after the execution of the science demonstration were in consonance with the students' predictions of the outcome of the science demonstration. Namely, a significant number of the students who participated in this science demonstration predicted that the heavier object would fall faster. Overall, these researchers offered that additional science demonstrations should be designed to aid in the discrediting of the students' pre-instructional conceptions; however, it was later observed that students were unwilling to alter these pre-instructional conceptions (Champagne, Gunstone, & Klopfer, 1985).

As a point of departure from the Champagne, Klopfer, and Anderson study, Gunstone and White (1981) investigated the understanding of gravity in first-year university students enrolled in a large enrollment physics course. Eight physical science scenarios related to gravity were presented to the students; this was done at the beginning of the academic term to insure that they would not be influenced by the lecture instruction. Each student was assigned to one of five groups. Since each group could only meet with the researchers for one hour, and data pertaining to all eight of the scenarios was to be procured, all of the scenarios were not experienced by all of the groups. For all of the physical science scenarios, each group of students was first introduced to the situations. The investigation followed the procedure of prediction, demonstration, observation, and explanation. The results of this study led the researchers to conclude that these students understood a great number of physics concepts; however, they did not relate them to their lives. They offered that science educators must focus their attention more on helping students negotiate school and general knowledge. In addition, there was an inability among the students to effectively explain their predictions; another outcome was that there was a penchant for each student to observe his or her own predictions. Similar findings were reported by Hynd, McWhorter, Phares, & Suttles (1994). McCloskey (1983) found that there was a discrepancy between students' intuition and the physical reality of the demonstration.

In 1986, O'Brien and Heikkinen, as reported by O'Brien (1988), investigated the effect that an Institute for Chemical Education (ICE) demonstration workshop had on a group of elementary, middle, and secondary school teachers. The two-week, intense training program was highly focused and skills-oriented, and placed great emphasis upon the four components of training effectiveness (e.g., diagnostic/prescriptive, presentation of theory/concept, modeling or demonstrating, and practice under simulated conditions with feedback components). The main goal of the program was to provide an intervention that would increase science teachers' use of chemical demonstrations. Participation in this program was voluntary. Prior to the program, the participants completed a questionnaire that identified their stages of concern about the innovation. Specifically, the questionnaire ascertained scores related to specific concerns related to self (e.g., awareness, informational, personal), task (e.g., management), and impact (e.g., consequence, collaboration, refocusing). Students were asked to complete the

35-Likert items (0-7 intensity scale, where 0 is low and 7 is high) keyed to the seven stages of concern related to self, task, and impact.

At the end of the program, the teachers completed the same questionnaire that was completed by the students. Post-program scores were compared to pre-programs scores. The results of the study indicated that the ICE demonstration workshop mediated a reduction in concerns that pertained to self, and simultaneously augmented the participants' scores associated with impact. Post-program scores were also reported for importance of lack of ideas, skills, confidence, safety, and classroom management issues. Also, the teachers' confidence level rose as a result of this intervention. However, there were no significant changes noticed by the researcher in those participants' scores related to management concerns (e.g. limitations of facilities, curriculum time, preparation/cleanup time, equipment/supplies, personal inertial). O'Brien (1988) reported similar findings in an investigation with another group of teachers.

The following research findings posed by Shepardson, Moje, and Kennard-McClelland (1994) also have direct implications for our research and is included here. Their study consisted of three consecutive demonstrations in which children were asked to make predictions, observations, and explanations of the phenomena being displayed. The participants in this study were fifty-two fifth-grade students from two separate classes. The teachers from both of these classes worked as a team during the presentation of the demonstrations.

For the first demonstration, the teachers placed a balloon over the mouth of flask; then, a flame was held beneath the flask. Next, the teachers placed a flame beneath an open paper bag that was secured to one end of a balance. The demonstration was utilized to convey to the students that air rises and expands when heated. After students witnessed these two demonstrations, the teachers then presented the students with the egg-in-the-bottle scenario. For this particular demonstration, "[t]he fire heats the air, causing the air to expand, rise, and leave the bottle. When the fire goes out, the remaining air in the bottle cools and contracts, creating a partial vacuum inside the bottle. The pressure inside the bottle is less than outside; thus, the egg is pushed in as air moves back into the bottle" (p. 257).

Seven children from the total of fifty-two who witnessed the demonstrations were interviewed. The results indicated that the children's predictions were influenced by "the emphasis on heat and rising hot air from the first two demonstrations" (p. 253), and the students' prior knowledge and personal experiences with fire and heat. Because a cause-effect relationship was not directly observable between the heat and the egg, these children indicated what they "expected, rather than what they observed" (p. 253). Despite the fact that one-third of all of these elementary students possessed a partial understanding of the concept of air pressure, of the seven who were interviewed, none were able to relate the egg-in-the-bottle demonstration to the concept of weather.

Silberman (1983) conducted a study where he performed a demonstration of a chemical phenomenon and asked his second semester general chemistry university students to describe and explain the chemical phenomenon. The specific demonstration was

> A 4-L beaker containing 3.5 L of distilled water and large magnetic stirring bar was stirred at the rate that caused a smooth vortex to form. Next, 35 ml of 0.1 M mercuric nitrate was added with enough 0.001 M nitric acid to prevent the formation of hydroxide and/or carbonate precipitates. Small portions of 2 M potassium iodide solution were added in the center vortex causing a precipitate of HgI_2 in the form of an "orange tornado" that gradually dissipated as stirring continued. The tornado can be formed or dispersed a few times until there is enough iodide in the system to cause a mercuric iodide precipitate to disperse throughout the system. At this point, KI is added at the perimeter of the beaker. Eventually the HgI_2 will dissolve, with its tornado disappearing last, as the I^- concentration increases enough to form a mercuric iodide complex that is soluble and almost colorless. (p. 996)

It is important to note that the researcher offered this assignment as an exam bonus. Also, the students were encouraged to research this demonstration with the aid of textbooks and library resources. The researcher stressed that the students should work independently, and that no consultation with another individual was allowed.

Of the seventy responses that were submitted, only one was correct and ten were partially correct. The remaining sixty-nine responses were "either totally wrong or showed major errors in understanding, comprehension, and/or reasoning" (p. 996). Some students utilized the concept names inappropriately, showing little or no understanding of the concept. For instance, he noticed that almost all of the terminology discussed in the lecture appeared in the student responses. There was evidence that the students contradicted themselves in their responses. One student wrote "HNO_3 acts as an acid" (p. 996); however, later that same student indicated that "HNO_3 is basic" (p. 996). Also, the students' "explanations showed an inability to consider a new approach or to explain a phenomenon that they had not seen before" (p. 996). The researcher noted that the students distorted chemical concepts and invented their own chemical equations to aid in the explanation of the phenomena observed in the demonstration. He was lead to believe that "[t]here was apparently an almost total unwillingness or inability (I'm not sure which) to attempt to discover what occurred by sorting out the relevant material in textbook or other reference in the library" (p. 996). Finally, he indicated that the students' responses suggested that they were unfamiliar with the properties of chemicals.

Silberman (1993) advised that even though the students could not describe and explain the events associated with the demonstration of a physical phenomenon, this was not justification for removing them from the curriculum. However, he added that, "it points to the need to do more and better demonstrations and to question students' understanding in order to improve student observational and interpretive skills" (p. 997). This could be accomplished, he recommended, by restructuring courses in such a way as to accommodate student discussions of observations as they pertained to demonstrations and experiments; this restructuring of the course also required that students be given opportunities to describe and explain observations on exam questions, instead of only relying on multiple choice tests and short answer exams.

The findings from Miller (1993) were also relevant to this research. Miller taught reduction and oxidation to his students in the form of a teacher demonstration. This demonstration also appears in Alyea & Dutton (1965). He placed eight grams of potassium hydroxide in 300 milliliters of water, all of which was housed in a 500-milliliter flask. The solution was allowed to cool; then, ten grams of dextrose was dissolved in the solution. A few milliliters of methylene blue solution were added; after a few minutes, the solution turned colorless. The science educator then shook the bottle, and the solution turned blue. After letting the solution rest, it returned to its initial colorless state.

The procedure for administering the demonstration followed a demonstration-exploration-discussion sequence. This sequence was also included as part of this science educator's student evaluation process. During the evaluation process, students were asked to record their observations, and to provide conclusions for the class demonstrations. Additionally, students were asked to respond to multiple open-ended questions, to describe their approach in resolving a problem, or to construct a short essay on various topics. He observed that students who received chemistry instruction via the demonstration-exploration-discussion method received higher achievement scores on the ACS (American Chemical Society) examinations as compared to those students who received instruction in a traditional lecture format. Also, he highlighted that there were multiple sections of this general chemistry course at this particular university. As compared with the students taught via the demonstration-exploration-discussion method, those students in the parallel course, taught by a different instructor, performed as well as "those students who attend classical lectures where each topic is more explicitly covered by the teacher" (p. 188).

Hur (1996) created a series of demonstration materials that were utilized to teach chemistry concepts in a large-lecture, university chemistry course; the participants in this study were science and engineering majors. In addition, a survey was administered to ascertain the students' attitudes toward the lecture demonstrations and examinations; pre—and post-tests were administered to procure students' understanding of the lecture demonstration-related chemistry concepts. Interviews, in-class writing assignments, and classroom observations were other sources of data. This data was collected over the course of a two-year period.

As a product of Hur's research, an instructional manual comprised of overhead projector lecture demonstrations was developed, and lecture demonstration descriptions, materials, formulae, data and equations accompanied each activity. The results of this study suggested that students found the overhead projector lecture demonstrations to be interesting and informative, provided that they entertained only one concept at a time. Generally, students performed better on the post-test after viewing the lecture-demonstrations. Overall, these students felt that the use of demonstrations during the lecture period enhanced their learning of the related concepts.

Also during the same year, Waldman, Schechinger, and Nowick (1996) described the results of their coordinated University of California-Irvine Chemistry Outreach Program for high school students. In this article, the authors discuss how the program evolved from 1991 to 1993 and the strategies that they utilized to attract undergraduate and graduate student volunteers to participate in the program. In this manuscript, the authors delineate the four types of shows (The Variety Show, The Variety Show II, The Polymer Show, The Organic Show) and related demonstrations that were scripted for viewing by high school students. Also during the show, the volunteers were asked to: (a) include the high school students during the demonstrations, personalizing the experience by using the student's name;

(b) discuss the volunteer's study of chemistry in high school, the path he/she took to study chemistry in college, and his/her typical day in graduate school; (c) distribute a sheet of activities that the students could perform at home to follow-up on their experiences from the outreach program; and (d) to perform an additional, follow-up experiment one-week post performance in their respective classrooms with their teachers. The authors related some interesting comments made by the volunteers who taught the program to high school students, as well as comments by the high school students and their teachers. Overall, the authors concluded that the outreach program benefited the volunteers and the high school students. Specifically, the volunteers gained an enhanced appreciation of teaching and of chemistry education, while the high school students were afforded opportunities to see live performances of chemistry demonstrations covering fascinating topics that generally were not performed in their classrooms.

Another type of chemistry outreach program designed for students in grades 7-12 from rural areas and towns was described by Lopez-Garriga, Munoz-Sola, Torres, et al. (1997). This program, entitled Science on Wheels, originated at the University of Puerto Rico and targeted schools that had limited resources and little access to hands-on experiences in science education. In brief, undergraduate and/or engineering students performed the science demonstrations. In doing so, they exposed the pre-college students to various science phenomena (combustion and the fire triangle, acid-base reactions and indicators, polymers, phase changes). The performance times for demonstrations ranged from 1 to 1.5 hours in duration. After the presentations, the pre-college students formed small groups around the student who performed the demonstrations, posing questions and inquiring about the programs offered at the university. These gatherings, as reported by the authors, were important in that the pre-college students' observations and concepts learned during the performances could be reinforced by the performers.

The workshops afforded the teachers opportunities to develop their own science skills through the use of hands-on experiences for later use as demonstrations in the classroom and/or projects to be completed by the students in their classrooms. The teachers were supervised during the workshops. Primarily, this was done to ensure that the teachers did not present demonstrations without science preparation. The authors noted that lack of preparation could be unsafe, environmentally dangerous, and misleading in terms of message and purpose of the demonstrations. Overall, the teachers learned how to prepare their own interesting and safe demonstrations from materials obtained from local stores (drug stores, hardware stores, supermarkets). Typically, 14 to 20 teachers, working in pairs, participated in the workshop. The duration of the workshop was 5 hours. The undergraduate and engineering students also participated as trainers for the teachers enrolled in the workshops.

The authors described the aggregate results from three years of data collected from the students and teachers regarding their perceptions of the science demonstrations and workshops, respectively. In summary, the results of their study revealed that science demonstrations could increase students' science literacy, even those who would not pursue science careers after high school. In addition, they recommended that pre-college science classes incorporate a mix of experiences imbued with demonstrations and hands-on activities. They also recommended that a teachers' participation in a workshop like the Science on Wheels program could help the teachers improve their students' performance and motivation in their science classes.

Roth, McRobbie, Lucas, & Boutonne (1997) investigated reasons that senior-year physics students failed to learn from demonstrations. They reported that

(a) Students have difficulties separating signals from noise—that is, they do not know which aspects of the display they need to focus on to understand the teacher's accompanying or following theory talk;

(b) when students come to see a particular demonstration, they bring with them different discourses that frame their descriptions and explanations, which may be inappropriate for and even interfere with the development of a discourse suitable for the situation at hand; (c) other demonstrations students have seen may interfere with their development of a discourse because of superficial similarities in images and discourse;

(d) students may not be able to connect the different representations that are implicit in the teacher's theory talk to other aspects of their knowledge about physical systems;

(e) low priority given to constructing and understanding phenomena compared to being able to get the correct results on numerical tasks affects students' engagement with the demonstrations; and

(f) a lack of opportunity for students to engage in a discourse about the demonstration to test the appropriateness and suitability for describing, constructing, and explaining phenomena. (pp. 520-521)

With these ideas in mind, they recommended that teachers modify their lecture-demonstration procedures to:

(a) Engage students in talking about and representing phenomena;
(b) Engage students in discussion about scientific inquiry and the construction of variables such as to produce a consistent theoretical framework;
(c) Construction of variables that allow them to keep account of systems despite change;
(d) Engage students in discussion about the mutually constitutive function of the language game and phenomenon, situated language, and knowledge that assists in the separation of signal from noise;
(e) Have students generate evidence and theory, set up a forum in which these are hammered out, and decide on future evidence to be needed and constructed. (p. 529)

In an ethnographic study, Candela (1997) found that when teachers showed experiments and exercises in their science classrooms, they frequently converted exercises explicated in textbooks into various demonstrations. The sample size was not reported in this particular article. Also, it was observed that these teachers converted demonstrations into problem-solving activities for the students. He indicated that the proposed curriculum and the constructed curriculum could be transformed by the student's observations and questions. This transformation, he added, "gives new significance to knowledge" (p. 510). Finally, from this study, he concluded,

> Demonstrations or proposed problems in didactic material can be transformed when they enter the classroom and can promote interaction among participants of the educational process. The public process of education resignifies the chores and applies new meanings to the knowledge proposed in textbooks and school programs." (p. 511)

In another study, Hatcher-Skeers and Aragon (2002) described their research with college students enrolled in a general chemistry course where active learning in the classroom was combined with service learning in the community. Working in small groups (4 or 5 students), the students worked in conjunction with the professor of the course to select and present science demonstrations for Chemistry Day, an event for middle school students.

From a list of resources, each group of college students was asked to select a demonstration that could be presented safely and that would be relevant to the content of their college chemistry course. In preparation for the middle school event, each group of college students was asked to present and explain their demonstrations, while fielding any questions from their peers. On Chemistry Day, the groups of college students presented their demonstrations to groups of middle school students and their respective teachers. As part of the presentation, the college students performed their demonstrations, described the chemistry concepts associated with the phenomena displayed, and answered questions about science and about their roles as college students. On course evaluations, the students were asked to share their feelings about their involvement with the demonstration-related community project. When asked if they liked preparing and presenting their own demonstrations and watching their peers perform demonstrations, 87% of the students responded positively. The researchers noted that the demonstrations often ranked higher than any other component of the chemistry course. However, most college students reported that they preferred doing the demonstrations themselves to having them performed by the professor. Students also believed that performing the demonstrations forced them to learn the related science concepts and helped in their overall studying. In addition, many students felt that the demonstration preparation gave them "renewed confidence in their laboratory skills" (p. 463). The researchers reported that this project "demonstrated that it is possible to incorporate presentations early in the students' college career and that they can be done even in large introductory classes" (p. 463). Although the students' participation in Chemistry Day was not mandatory, the researchers noted that 75-80% of students enrolled in the course participated voluntarily. They ascribed the high volunteer rate to the popularity and success of the student-driven demonstration project.

Majerich (2004) investigated the development of understandings of chemistry topics for university students enrolled in a large-enrollment chemistry non-laboratory, lecture demonstration course for non-majors. Ascertained in this study were: (a) perceptions of the benefits of using science lecture demonstrations and related discussions; (b) students' understandings of chemistry as reported on related assessments thereof; (c) differences in levels of understanding among students; and (d) the extent that students' understandings of chemistry indicated that meaningful learning had occurred. The participants were 171 undergraduate students, ranging from sophomore to junior level. Fourteen students had previous experience with a chemistry course at the high school level; none of these students had experience with a university or college chemistry course.

The course was predicated upon the teaching of chemistry-related concepts through discussions of demonstrations. In total, the students experienced and discussed eighty-nine demonstrations over course of the semester. The topics discussed in the class were Chemical Change, Physical Change, Chemical Reactions, Density, the Laws of Chemistry, Decomposition, Periodic Table, Cathode Ray Tubes, Radioactivity, Electromagnetic Radiation, and Atomic Structure. The overarching topic that shaped the course was chemical reaction, in that what the students learned would better help them identify what was or was not a chemical reaction.

Due in part to the large enrollment and limited availability of science laboratories, the course used a participatory science lecture demonstration (SLD) technique. The SLD method required that the professor and students proceed through the demonstrations while relating, through discussion, each science demonstration to its predecessor. In addition, at the beginning of each class, a quiz was administered to determine the extent of the students' understanding of chemistry developed from the previous demonstrations and related discussions. The quizzes were reviewed with the students immediately upon completion, and this served as the starting point for that day's classroom discussions. During weeks 6, 9, and 15 of the fifteen-week course, the students were given an assessment. Both quiz and assessment items required that the students: (a) delineate science terminology; (b) recall science lecture demonstrations; and (c) activate science knowledge. In addition, two anonymous, student-completed course evaluations were administered during weeks 6 and 15, respectively. Students were asked to provide responses to: (a) What are the benefits associated with the science lecture demonstrations used in this course?' and (b) Why are science lecture demonstrations being used in this course?"

For the majority of topics discussed in the course, students were more capable of delineating terms than recalling science demonstrations and activating their knowledge. However, the students were better able to recall science demonstrations than activate the science knowledge learned from the demonstrations. When the students' responses to assessment items were further parceled into one of three levels of understanding of chemistry (novice, transitional, and shared—with the professor), it was concluded that the students who developed shared understandings of the core science topics (Chemical Property, Physical Property, Chemical Change, Physical Change, Chemical Reactions) were more successful at developing shared understandings of related topics (Decomposition, Laws of Chemistry, The Periodic Table) later in the semester.

Most of the students who developed transitional or novice understandings of the core science topics (Chemical Property, Physical Property, Chemical Change, Physical Change, Chemical Reactions) were less successful at developing shared understandings of related topics (Decomposition, Laws of Chemistry, The Periodic Table) later in the semester. However, these underdeveloped understandings of the early topics did not necessarily hinder the development of more complete understandings of topics (Cathode Ray Tubes, Radioactivity/Electromagnetic Radiation, Atomic Structure) discussed later in the course.

Evidence for the students' meaningful learning of science knowledge was ascertained on items requiring: (a) the extension of science learned when applied to novel situations; (b) generalizable statements made about science scenarios; and (c) the differentiation between the meanings of terms. Results showed that for the majority of topics (Chemical Property, Physical Property, Chemical Change, Physical Change, Chemical Reactions, Decomposition, Laws of Chemistry, The Periodic Table), more students consistently had understandings of these topics shared by others than for topics (Cathode Ray Tubes, Radioactivity/Electromagnetic Radiation, Atomic Structure) developed later in the semester. This is attributed to the fact that the science lecture demonstrations were better interrelated early on in the semester; however, this was not always the case with discussions of science lecture demonstrations of unrelated science topics introduced near the end of the semester.

The students' responses on the open-ended surveys revealed that over half of the students' perceptions of the benefits of science lecture demonstrations were linked to a feature of the students' learning of chemistry material (inculcation of science process skills, science knowledge gained, enhancement of previously learned science knowledge, confluence of materials/textbook/ discussion, and/or recollection of demonstration-related discussions). Over one-sixth of the students' perceptions were linked to a contribution made in the development of their understandings of chemistry (modeled how to extend science knowledge to new situations, used science terminology properly, and/or applied science knowledge). Also, over one-sixth of all students' perceptions of demonstrations were linked to their disposition in learning and developing understandings of chemistry (offered a new perspective to view science phenomena, showing that chemistry is a human endeavor, increased their appreciation for chemistry, emphasized the importance of chemical knowledge, increased their confidence to study chemistry, stimulated their interest in chemistry, highlighted their responsibility for learning the course material, and/or revealed the importance of the professor/student interaction in discussions.

Ophardt, Applebee and Losey (2005) investigated how non-major students felt about demonstration-focused laboratory component of a university chemistry course compared to other science courses (biology, physics, and geology) with traditional laboratories offered from other science disciplines at the non-major level. In addition, the researchers explored the extent that student-learning outcomes were met in this course that was taught for two years with two classes per year. The class met for 65-minute lecture sessions and one three-hour laboratory meeting. When in the laboratory, each student was responsible for selecting two demonstrations to prepare and perform for that lab class and later at a demonstration show. The students worked in the laboratory to create and practice their demonstrations, which were related to the topics discussed in class. The instructor suggested demonstrations to those students who were struggling with the course chemistry concepts. In addition, the instructor met with each of the students to assist each student with concepts and molecular-level explanations, as well as to critique and provide presentation and/or performance ideas to the students. After three weeks of demonstration preparation, the students were assigned to smaller groups (4 or 5 students), and they performed a dress rehearsal of the show at their college. Each student performed four demonstrations with 20-24 demonstrations presented during any given show. The instructor used a rubric to score each student's performance with respect to: (a) introductory comments; (b) commentary during the demonstration; (c) explanation; and (d) demonstration skill and poise. During the semester, the entire class performed demonstration shows twice for students at a local elementary school. Using a questionnaire, students were asked to rank six qualities of the chemistry laboratory experience: (a) learning basic chemistry concepts; (b) improving oral communication skills; (c) improving written communication skills; (d) making connections between chemical theory and concrete examples; (e) overcoming your fears of chemicals; and (f) strengthening you interest in science. Results showed that the demonstration-based laboratory was effective in reaching course goals for students. Students were also asked to compare their demonstration-based laboratory experiences with their experiences with traditional laboratories on four qualities: (a) interest in activities; (b) level of understanding of laboratory material; (c) confidence in working in the laboratory; and (d) freedom to learn at your own pace. Results indicated that most students preferred the demonstration-based laboratory to the traditional laboratory. The researchers concluded that the demonstration-based laboratory course in combination with the outreach program had numerous benefits for students in the course and for the instructor who taught it. During the laboratory, the students were afforded opportunities to participate in "hands-on" experiences prior to the lecture on the related topics. In working with students, the laboratory session afforded the instructor the opportunity to pose questions to the students on the theory related to the demonstration because they had observed a concrete example, which is the substrate for an explanation. In addition, the instructor noted an increase in students' interest in the course material, a willingness to ask more questions, and an improved understanding of the scientific principles surrounding the demonstrations. In summary, the researchers added "[t]his course demonstrates the beauty and commonality of chemistry in everyday life, and it alleviates the fear of "chemicals" among this group of nonscience friends" (p. 1177).

In their preliminary work, Milner-Bolotin, Kotlicki, and Rieger (2007) confirmed that students in a large, introductory, college, physics class, who were shown instructor-centered demonstrations "remembered not what they saw, but what they expected to see" (p. 46). In order to help students develop improved conceptual understandings of the course content, the authors designed and implemented a five-stage technique for presenting lecture demonstrations, known as Interactive Lecture Experiments (ILEs). The technique was modeled after the Activity-Based Physics Groups' Interactive Lecture Demonstration (ILD) technique. While both techniques engage students in making and discussing predictions prior to observing the demonstration, the ILE technique was developed to avoid creating the "negative emotional impact of cognitive conflict on students' confidence (p. 47).

The stages of the ILE include: (a) students view the demonstration set-up and the instructor provides an overview of related concepts; (b) students observe the demonstration and gather data; (c) students cooperatively analyze the data outside of class; (d) the instructor facilitates a student-centered discussion of the data analysis; and (e) students apply the new concepts in problem solving activities that begin in-class and extend as a homework assignment. Four ILEs were implemented with 750 students by the authors during the fall semester of 2005. Over that time period the students' perception of the usefulness of the ILE technique shifted positively from 52% to 78%. Additionally, the researchers reported a significant improvement on exam questions compared to the previous semester in which only the traditional demonstrations were utilized.

Expanding upon earlier research (Majerich, 2004, described above), Majerich and Schmuckler (2007) provide an in-depth comparison of students' perceptions of demonstrations and content mastery as a result of exposure to the non-traditional, science lecture demonstration (SLD) teaching method or the traditional lecture demonstration (TLD) teaching method in a large-enrollment, chemistry lecture demonstration course for nonscience majors.

The SLD method was designed to improve upon the use of lecture demonstrations to promote greater content mastery and student engagement. The SLD method possesses several unique aspects to achieve this objective. The instructor encourages direct student involvement in performing the demonstrations when appropriate. The instructor also poses convergent and divergent questions related to each demonstration in order to facilitate student-centered discussions. SLD students received "additional strategies for observing demonstrations and organizing their class notes during class, as well as for revisiting their notes about demonstrations outside of class" (p. 63).

In order to assist students with their observation skills, the authors injected elements of psychological research, specifically the concept of *inattentional blindness* (Mack & Rock, 1998), into the modified SLD teaching method. Inattentional blindness is described as "the failure on behalf of students to perceive the dynamic event (demonstration). "To compensate for this, SLD students were asked to focus their attention on a highly specific region (referred to here as the *window of observation*) of each demonstration" (p. 63). For each of the 102 demonstrations, the instructor identified a window of observation. As an additional strategy, the instructor encouraged students to relate each demonstration and its discussion to the previous one shown in order to build coherency throughout the course content.

To assess content mastery in the TLD section, the instructor administered examinations containing a multiple-choice section followed by a section of open-ended questions. The SLD section's examinations consisted of only open-ended items. Both examination types included questions structured to evaluate recall of demonstrations and the application of new chemistry knowledge. Both sections' examinations covered similar content, but statistical differences could not be directly determined due to the format difference. However, the authors note with caution that the SLD students consistently received higher examination scores.

At the conclusion of the course students in both sections completed a survey assessing perceived benefits of the use of science demonstrations. The survey items addressed each of the 10 merits associated with the teaching and learning of science using demonstrations (Majerich, 2004). Analyses revealed significant differences between the TLD and SLD students on five of the ten merits. Students in the SLD section perceived more often than students in the TLD section that the use of demonstrations: (a) showed abstract concepts with the use of concrete examples; (b) enhanced learning of course material; (c) served as a substitution for laboratory exercises; (d) developed creativity and a sense of cooperation; and (e) revealed the professor's attitude toward chemistry. Course statistics also reveal that a smaller percentage of students in the SLD section (3.3%) withdrew from the course for academic reasons than the students in the TLD section (22%).

In summary, the authors recommend that science instructors explicitly encourage student discussion of demonstrations with the instructor and with each other both in and out of class. At the same time instructors should better connect and sequence the demonstrations utilized in the classroom. In order to promote increased content mastery, instructors must also offer students strategies for observing and recalling demonstrations.

Building upon previous work (Majerich, 2004; Majerich & Schmuckler, 2007; Majerich, Fadigan, & Schmuckler, 2008), Majerich & Schmuckler (2008) provide an in-depth explanation and detailed procedure for implementing the indigo carmine oxidation-reduction reaction demonstration in a large lecture hall format. Their methodology enables students to begin to "develop and describe several of the concepts associated with these chemical reactions" so that students are constructing their own conceptual understanding before being exposed to the technical details of the demonstration (p. 14).

Through observations students gather qualitative data about the phenomenon such as color change, reaction rates, reversible reactions, energy requirements, and equilibrium. The lecturer's highly specific instructions to students on *when* to make observations throughout the demonstration, combined with explicit instructions on *where* to focus their attention, i.e. the "window of observation" (Majerich & Schmuckler, 2007, p. 63), increase student engagement in the demonstration process. Students are encouraged to make multiple observations of the phenomenon during the class session, and then again revisit the phenomena in the next class session to observe any additional changes that have occurred.

Overall, the authors provide a specific and practical example of how to incorporate the four considerations for successful teaching with demonstrations (Majerich, Fadigan, & Schmuckler, 2008). Their example incorporates: (a) student discussion; (b) data collection (recording observations); (c) compensation for inattentional blindness through multiple opportunities to view the phenomenon; and (d) identifying a "window of observation" (Majerich & Schmuckler, 2007, p. 63). Although the article addresses chemistry teaching, the processes can very well apply to other areas of science teaching.

SUMMARY

The purpose of this compendium was to review the research that pertained to the use of demonstrations for the teaching of science from 1918 to 2008. As research studies were located, each investigation was chronologically organized and partitioned into one of three overarching themes assigned to it by these researchers. The studies were grouped according to comparison themes, i.e., comparing the lecture demonstration method and the individual laboratory method of teaching science (1918-1989), further types of demonstration-related comparison studies (1958-2008), and single demonstration studies (1980-2008). Organization of the research in this manner revealed that an interest in investigating this type of research by science educators had and continues to evolve.

Initially, there were studies comparing the effectiveness of the methods of lecture demonstration and the individual laboratory. The very early attempts (1918-1964) to determine the superiority of the lecture demonstration method over the individual laboratory method of teaching science, and vice versa, remained unresolved through the mid-1960s. Cunningham (1946) suggested several reasons for the inconclusive nature of the results reported by researchers of lecture demonstration and individual laboratory comparison studies. After carefully reviewing the studies up to 1946, he noted that there were numerous variables that should have remained fixed for the duration of the experiments. Some of the variables that confounded the results of those investigations included the uncontrolled variables related to: (a) the teacher; (b) the complexity of experiments and apparatus; (c) the time spent on each method; (d) the amount of science studied by students; and (e) the performer of the demonstrations.

Even after exacting a procedure to control for the teacher variable reported by Cunningham (1946), Yager et al. (1969) were able to show that students who participated in a discussion-demonstration group or discussion-laboratory group developed more skills than those students who received science instruction via a discussion-only method. The results of Yager and his predecessors suggested that neither the demonstration method nor the laboratory method for the teaching and learning of science was superior.

From 1958 to 2008, investigations comprised of multiple demonstration-related comparison studies were also unable to confirm the superiority of the demonstration method for the teaching and learning of science as compared to other methods in each study. For instance, Oliver (1975) compared three methods of teaching high school biology—lecture-discussion, a combination of lecture-discussion and demonstration, and demonstrations. Initially, although he determined that the lecture-discussion method of teaching biology was noticeably superior in terms of biology content acquisition during the onset of the semester, this result was ephemeral, and dissipated at the termination of the following semester. Comparing a self-paced instructional method and a teacher demonstration method of teaching college chemistry, Eniaiyeju (1983) concluded that students' achievement scores were higher when they participated in the self-paced program; in addition, most of the students actually preferred the self-paced programs. Kraus (1997) abandoned her efforts to determine the effectiveness of the demonstration method when she noticed that students were unable to obtain a functional understanding of the demonstration-related physics concepts.

Recently, some research investigations have focused on an isolated group of students and the demonstration method of teaching and learning of science (Fagen, 2003; Kraus, 1997; Majerich, 2004; Majerich & Schmuckler, 2007; Ophardt, Applebee & Losey, 2005). For instance, Roth et al. (1997) focused their attention on one specific group of senior-year physics students in an effort to explore in-depth the effectiveness of demonstrations and their relationship to student learning. Overall, they offered six reasons why these students failed to learn from exposure to demonstrations. Clearly, the research of Roth et al. offered new insight into the difficulties associated with the learning of science via the demonstration method. The work of Majerich (2004), Majerich and Schmuckler (2006; 2007; 2008), and Majerich, Fadigan, & Schmuckler (2008) also shed new insight into how to better prepare students to learn from demonstrations as an instructor-centered, traditional method of teaching was replaced with a more student-centered, science lecture demonstration method.

As informed science teachers, we need to remain cognizant of issues and trends, new and old, discussed in the science education (scholarly) literature and existing (empirical) research. We have shown that the literature and research may not always be synchronized, so science teachers need to view the discrepancy from a new perspective. By not preparing students to participate in the inquiry process during demonstrations, the perceiving of the science phenomenon, the learning of science topics, the developing of understandings of science may not be achieved.

This compendium highlights a ninety-year history of science education focused on how demonstrations have been used for the express teaching and learning of science at the high school, college and university levels. For those who do not engage in an examination of this past, the authors pass along Santayana's (1905) caution to those who do not study history in general: "Those who cannot remember the past are condemned to repeat it" (p. 284). In other words, science educators are relegated to repeat and confirm the poor performing student learning outcomes of the past if they remain uninformed about past practices. This compendium is a reference that includes what has been practiced in the past. The authors offer the compendium as a resource guide for science educators who wish to use demonstrations while learning from the past.

REFERENCES

Alyea, H. N., & Dutton, F. B. (1965). *Tested demonstrations in chemistry.* Easton, PA: Journal of Chemical Education.

Anibel, F. G. (1923). *Comparative effectiveness of the lecture-demonstration and individual method.* Unpublished master's thesis, University of Chicago, Chicago, IL.

Anibel, F. G. (1924). Comparative effectiveness of the lecture-demonstration and individual laboratory method. *Journal of Educational Research, 25,* 371-379.

Anibel, F. G. (1926). Comparative effectiveness of the lecture-demonstration and individual laboratory method. *Journal of Education Research, 13,* 355-365.

Ausubel, D. (1963). *The Psychology of Meaningful Verbal Learning.* New York: Grune & Stratton.

Ausubel, D. P. (2000). *The acquisition and retention of knowledge: A cognitive view.* Boston: Kluwer Academic Press.

Bates, G. C. (1978). Learning in science laboratories. In M. B. Rowe (Ed.), *What research says to the science teacher: Vol. 1* (pp. 55-82). Washington, DC: NSTA.

Beall, H. (1996). Report on the WPI conference "Demonstrations as a Teaching Tool in Chemistry: Pro and Con." *Journal of Chemical Education, 73*(7), 641-642.

Beasley, W. (1982). Teacher demonstrations: The effect on student task involvement. *Journal of Chemical Education, 59*(9), 789-780.

Bonwell, C.C., & Eison, J. A. (1991). *Active learning: Creating excitement in the classroom. ERIC Clearinghouse on Higher Education.* Washington, DC: The George Washington University, School of Education and Human Development. (ERIC Document Reproduction Service Nos. ED336049 and ED340272).

Boretz, N. (1930). *Individual experiment versus pupil demonstration methods in high school general science.* Unpublished master's thesis, New York University, New York.

Bowen, C. W., & Phelps, A. J. (1997). Demonstration-based cooperative testing in general chemistry: A broader assessment-of-learning technique. *Journal of Chemical Education, 74*(6), 715-719.

Bradley, R. L. (1965). Lecture demonstration versus individual laboratory work in a general education science course. *The Journal of Experimental Education, 34*(1), 33-42.

Brasure, R. E. (1929). An experimental study of the teacher demonstration and the individual laboratory methods in the teaching of physics. Unpublished master's thesis, University of Wisconsin, Madison, WI.

Buchanan, S. A., Reynolds, M. M., Duersch, B. S., Lohr, L. L., Coppola, B. P., Zusho, A., & Pintrich, P. R. (2004). Promoting student learning in a large general chemistry course. *Journal of College Science Teaching, 33*(7), 12-17.

Buncick, M. C., Betts, P. G., & Horgan, D. D. (2001). Using demonstrations as a contextual road map: Enhancing course continuity and promoting active engagement in introductory college physics. *International Journal of Science Education, 23*(12), 1237-1255.

Bybee, R. W. (1970). The effectiveness of an individualized approach to a general education earth science laboratory. *Science Education, 54*(2), 157-161.

Candela, A. (1997). Demonstrations and problem-solving exercises in school sciences: Their transformation within the Mexican elementary school classroom. *Science Education, 81*(5), 497-513.

Carpenter, W. W. (1925). Certain phases of the administration of high school chemistry. Unpublished doctoral dissertation, Teachers College, Columbia University, New York, NY.

Carpenter, W. W. (1926). A study of the comparisons of different methods of laboratory practice on the basis of results obtained on tests of certain classes in high school chemistry. *Journal of Chemical Education, 3*(7), 798-805.

Champagne, A. B., & Gunstone, R. F., & Klopfer, L. E. (1985). Instructional consequences of students' knowledge about physical phenomena. In L. H. T. West & W. L. Pines (Eds.), *Cognitive structure and conceptual change* (pp. 61-90). Orlando, FL: Academic Press.

Champagne, A. B., Klopfer, L. E., & Anderson, J. H. (1980). Factors influencing the learning of classical mechanics. *American Journal of Physics, 48*(12), 1074-1079.

Chiappetta, E. L., & Koballa, T. R. (2002). *Science instruction in the middle and secondary schools.* Upper Saddle River, NJ: Merrill Prentice-Hall.

Cooprider, J. L. (1922, December). Laboratory methods in high school science. *School Science and Mathematics*, 23, 526-530.

Cooprider, J. L. (1923). Oral versus written instruction and demonstration versus individual work in high school science. *School Science and Mathematics, 22,* 838-845.

Coulter, J. C. (1966). The effectiveness of inductive laboratory, inductive demonstration, and deductive laboratory in biology. *Journal of Research in Science Teaching, 4*(4), 185-186.

Crouch, C. H., Fagen, A. P., Callan, J. P., & Mazur, E. (2004), Classroom demonstrations: Learning tools or entertainment? *American Journal of Physics, 72*(6), 835-838.

Cunningham, H. A. (1920). Individual laboratory work vs. lecture demonstration. *University of Illinois Bulletin, 27,* 104-117.

Cunningham, H. A. (1946). Lecture demonstration versus individual laboratory method in science teaching—a summary. *Science Education, 30*(2), 70-82.

Cunningham, H. A. (1924, November). Laboratory methods in natural science teaching. *School Science and Mathematics, 24,* 848-851

Cunningham, H. A. (1924, October). Laboratory methods in natural science teaching. *School Science and Mathematics, 24,* 709-715.

Cunningham, H. A. (1924). Laboratory methods in natural science teaching, part I. *School Science and Mathematics, 24,* 709-715.

Cunningham, H. A. (1924). Laboratory methods in natural science teaching, part II. *School Science and Mathematics, 24,* 848-857.

Dearden, D. (1962). A study of contrasting methods in college general biology laboratory instruction. *Science Education, 46*(5), 399-401.

Deese, W. C., Ramsey, L. L., Wolcyzk, J., & Eddy, D. (2000). Using demonstration assessments to improve learning. *Journal of Chemical Education, 77*(11), 1511-1516.

DeJarnett, W. W. (1934). A study of the lecture-demonstration and individual laboratory methods of teaching biology in a small high school. Unpublished master's thesis, University of Kansas, Lawrence, KS.

Downing, E. R. (1924). A comparison of the lecture-demonstration and the laboratory methods of instruction in science. *School and Society, 19*(496), 769-770.

Downing, E. R. (1925, November). A comparison of the lecture-demonstration methods of instruction in science. *The School Review, 33,* 688-697.

Downing, E. R. (1927). Individual laboratory work versus teacher demonstration. *Science Education, 11*(2), 96-99

Downing, E. R. (1929, April). Shall laboratory work in the public school be curtailed—A criticism. *School Science and Mathematics, 29*, 411-413.

Downing, E. R. (1931). Methods in science teaching. *Journal of Higher Education, 2*(6), 316-320.

Downing, E. R. (1931). Methods in science teaching. *Journal of Higher Education, 2*(6), 316-320.

Dyer, J. H. (1927). *An analysis of certain outcomes on the teaching of physics in public schools with an investigation of the efficiency of a laboratory method in establishing such outcomes.* Unpublished doctoral dissertation, University of Pennsylvania, Philadelphia, PA.

Eniaiyeju, P. (1983). The comparative effects of teacher-demonstration and self-paced instruction on concept acquisition and problem-solving skills of college level chemistry students. *Journal of Research in Science Teaching, 20*(8), 795-801.

Erickson, H. A. (1929). *Laboratory experiments versus demonstration experiments in high school chemistry.* Unpublished master's thesis, New York University, New York, NY.

Ewing, E. (1926). *Individual laboratory versus teacher demonstration method of teaching chemistry.* Unpublished master's thesis, Temple University, Philadelphia, PA.

Fagen, A. P. (2003). Assessing and enhancing the introductory science course in physics and biology: Peer instruction, classroom demonstrations, and genetics vocabulary. *Dissertation Abstracts International, 64*(05), 1586A. (UMI No. 3091550).

Frings, H. & Hichar, J. K. (1958). An experimental study of laboratory teaching methods in general zoology. *Science Education, 42*, 255-261.

Garett, R. M., and Roberts, I. F. (1982). Demonstration versus small group practical work in science education. A critical review of studies since 1900. *Studies in Science Education 9*, 109-146.

Gattis, K. W. (1995). An investigation of the conceptual changes resulting from the use of demonstrations in college physics. *Dissertations Abstract International, 56*(04), 1303A. (UMI No. 9526926).

Glasson, G. E. (1986). The effect of hands-on versus teacher demonstration instructional methods on science achievement in relation to cognitive reasoning ability. *Dissertation Abstracts International, 47*(05), 1677A. (UMI No. 8618345).

Glasson, G. E. (1989). The effects of hands-on and teacher demonstration laboratory methods on science achievement in relation to reasoning ability and prior knowledge. *Journal of Research in Science Teaching, 26*(2), 121-129.

Goldstein, P. (1937). Developing laboratory resourcefulness. *Science Education, 21*(4), 185-193.

Goldstein, P. (1937). *Student laboratory work versus teacher demonstration as a means of developing laboratory resourcefulness.* Unpublished master's thesis, College of the City of New York, New York, NY.

Goldstein, P. (1937). Student laboratory work versus teacher demonstration as a means of developing laboratory resourcefulness. *Science Education, 21*(4), 185-193.

Gould, A. B. (1931). Demonstration experiments and their place in the teaching of chemistry. *Journal of Chemical Education, 8*(2), 297-302.

Gunstone, R. F., & White, R. (1981). Understanding of gravity. *Science Education, 65*(3), 291-300.

Gunstone, R. F., White, R. T., & Fensham, P. J. (1988). Developments in style and purpose of research on the learning of science. *Journal of Research in Science Teaching, 25*(7), 513-529.

Hatchers-Skeers, M. & Aragon, E. (2002). Combining active learning with service learning: A student-driven demonstration project. *Journal of Chemical Education, 79*(4), 462-464.

Hix, E. L. (1933). *The value of laboratory work in high school physics.* Unpublished master's thesis, State College of Washington, Pullman, WA.

Horton, R. E. (1928). *Measurable outcomes of individual laboratory work in high school chemistry.* (Teachers College, Columbia University, Contributions to Education, No. 303). New York: Bureau of Publications, Columbia University.

Horton, R. E. (1929). Does laboratory work belong? *Journal of Chemical Education, 6*(6), 1130-1135.

Horton, R. E. (1930). Measured outcomes of laboratory instruction. *Science Education, 14,* 311-319.

Hunter, G. W. (1922, January). Laboratory methods in natural science teaching. *School Science and Mathematics, 24,* 29-32.

Hunter, G. W. (1934). *Science teaching.* New York: American Book Company.

Hur, C. (1996). Development of demonstration-enhanced courses throughout the chemistry curriculum and a comparative study of student response. *Dissertation Abstracts International, 57*(03), 1083A. (UMI No. 960807).

Hurd, A. W. (1929). *Problems of science teaching at the college level.* Minnesota: University of Minnesota Press.

Hynd, C. R., McWhorter, J. Y., Phares, V. L., & Suttles, C. W. (1994). The role of instructional variables in conceptual change in high school physics topics. *Journal of Research in Science Teaching, 31*(9), 993-946.

Johnson, P. O. (1928, September). A comparison of the lecture-demonstration, group laboratory experimentation, and individual laboratory experimentation methods of teaching high school biology. *Journal of Educational Research, 18,* 103-111

Johnson, H. (1991). *The relative effectiveness and efficiency of hands-on, demonstration and videotape laboratories for non-science major students.* Unpublished master's thesis, Washington State University, Washington.

Kahn, P. (1937) An experimental study to compare the laboratory method of instruction with individual demonstration in elementary college biology. Unpublished master's thesis, The College of the City of New York, New York, NY.

Kiebler, E. W., & Woody, C. (1923). The individual laboratory method versus the demonstration method of teaching physics. *Journal of Education Research, 7,* 50-58.

Knox, W. W. (1926). *The demonstration method versus the laboratory method of teaching high school chemistry.* Unpublished master's thesis, University of Texas, Austin, TX.

Knox, W. W. (1927). The demonstration method versus the laboratory method of teaching high school chemistry. *The School Review, 35,* 376-386.

Knox, W. W. (1927, May). The demonstration method versus the laboratory method of teaching high school chemistry. *The School Review, 25,* 376-386.

Kraus, P. A. (1997). Promoting active learning in lecture-based courses: Demonstrations, tutorials, and interactive tutorial lectures. *Dissertation Abstracts International, 58*(06), 2143A. (UMI No. 9736313).

Lopez-Garriga, J., Munoz-Sola, Y., Torres, V., Echevarria, Y., Nazario, W., de Jesus-Bonilla, W., & Camach-Zapata, R. (1997). Science on wheels: A coherent link between education perspectives. *Journal of Chemical Education, 74*(11), 1346-1349.

Mack, A., & Rock, I. (1998). *Inattentional Blindness.* Cambridge, MA: MIT Press.

Majerich, D. M. (2004). Developing understandings of chemistry in a large-enrollment science lecture demonstration-based course for non-majors: The extent of meaningful learning and implications for practice. *Dissertation Abstracts International, 65* (03), 881A. (UMI No. 3125541).

Majerich, D. M., Fadigan, K., Schmuckler, J. S. (2008, January). *What is Past is Prologue: Four New Research-based Considerations for Preparing Students to Perceive and Learn from Science Demonstrations.* Paper presented at the Hawaii International Conference on Education, Honolulu, HI.

Majerich, D. M., & Schmuckler, J. S. (2008). Demonstrating indigo carmine oxidation-reduction reactions: A choreography for chemical reactions. *Journal of College Science Teaching, 37* (4), 14-16.

Majerich, D. M., & Schmuckler, J. S. (2007). Improving students' perceptions of benefits of science demonstrations and content mastery in a large-enrollment chemistry lecture demonstration course for nonscience majors. *Journal of College Science Teaching, 36* (6), 60-67.

Manaf, E. B. A., & Subramaniam, R. (2004). *Use of chemistry demonstrations to foster conceptual understanding and cooperative learning among students.* Paper presented at the conference of the International Association for the Study of Cooperation in Education, June 22-24, 2004, Singapore.

Mayman, J. E. (1912). *An experimental investigation of the book method, lecture method, and experimental method of teaching science in the elementary schools.* Unpublished doctoral dissertation, New York City, Bureau of Educational Research, New York.

McCloskey, M. (1983). Intuitive Physics. *Scientific American, 248*(4), 122-130.

Miller, T. L. (1993). Demonstration-exploration-discussion: Teaching chemistry with discovery and creativity. *Journal of Chemical Education, 70*(3), 187-189.

Milner-Bolotin, M., Kotlicki, A., & Rieger, G. (2007). Can students learn from lecture demonstrations? The role and place of interactive lecture experiments in large introductory science courses. *Journal of College Science Teaching, 36*(4), 45-49.

Mintzes, J. J., Wandersee, J. H., & Novak, J. D. (1998). *Teaching science for understanding: A human constructivist view.* San Diego: Academic Press.

Mintzes, J. J., Wandersee, J. H., & Novak, J. D. (2000). Assessing science understanding: A human constructivist view. San Diego: Academic Press.

Nash, H. B., & Phillips, M. J. W. (1927). A study of the relative value of the three methods of teaching high school chemistry. *Journal of Educational Research, 15*(5), 371-379.

Novak, B. J. (1960). Clarifying language in science education. *Science Education, 44*(4), 321-328.

O'Brien, T. (1988). *A concerns-based field study of a series of two-week, chemical demonstration inservice programs.* Paper presented at the 1987 Fall Conference of the American Educational Teachers Society-Northwestern Region, October 30.

O'Brien, T., & Heikkinen. H. (1987). *An exploratory, concerns-based field study of a two-week, summer inservice program to increase science teachers' use of chemical demonstrations.* Paper presented at the Annual Meeting of the National Association for Research in Science Teaching, Washington, DC, April 24.

Ogborn, J., Kress, G., Martins, I., & McGillicuddy, K. (1996). *Explaining science in the classroom*. Philadelphia: Open University Press.

Okpala, P., & Onocha, C. (1988). The relative effects of two instructional methods in students' perceived difficulty in learning physics concepts. *Kenya Journal of Education, 4*(1), 147-161.

Ola-Adeniyi, E. (1985). Misconceptions of selected ecological concepts held by some Nigerian students. *Journal of Biological Education, 19*, 311-316.

Oliver, M. (1975). The efficiency of three methods of teaching high school biology of lecture-discussion, lecture-discussion and demonstration, and lecture-discussion and demonstration in combination with laboratory exercises. *Journal of Experimental Education, 33*, 289-298.

Ophardt, C. E., Applebee, M. S., & Losey, E. N. (2005). Chemical demonstrations as the laboratory component in nonscience majors courses: An outreach-targeted approach. *Journal of Chemical Education, 82*(8), 1174-1177.

Payne, V. F. (1931). *The lecture-demonstration and individual laboratory methods compared: (1) The literature; (2) the distribution of time; and (3) experimental.* Unpublished doctoral dissertation, University of Kentucky, Lexington, KY.

Payne, V. F. (1931). The lecture-demonstration and individual laboratory methods compared: (1) The literature. *Journal of Chemical Education, 9*(5), 932-937.

Payne, V. F. (1931). The lecture-demonstration and individual laboratory methods compared: (2) the distribution of time. *Journal of Chemical Education, 9*(6), 1097-1102.

Payne, V. F. (1931). The lecture-demonstration and individual laboratory methods compared: (3) experimental. *Journal of Chemical Education, 9*(7), 1277-1294.

Peiper, J. W., & Sutman, F. X. (1970). A brief historical analysis of the demonstration in the teaching of biology. *Science Education, 54*(1), 83-86.

Phillips, T. D. (1920, June). A study of notebook and laboratory work as an effective aid in science teaching. *School Review, 28*, 451-453.

Pruitt, C. M. (1925). *An experiment on the relative efficiency of methods of conducting chemistry laboratory work.* Unpublished master's thesis, Indiana University, Bloomington, IN.

Pugh, D. B. (1929). Comparison of lecture demonstration and individual laboratory methods of performing chemistry experiments. *High School Teacher, 3*(9), 384-386.

Roth, W-M., McRobbie, C., Lucas, K. B., & Boutonne, S. (1997). Why may students fail to learn from demonstrations? A social practice perspective on learning in physics. *Journal of Research in Science Teaching, 34*(5), 509-533.

Santanya, G. (1905). Life of Reason: Reason in Common Sense (Vol. 2). New York: Charles Scribner's Sons.

Scott, V. G. (1929). *An experiment in the effect of the individual laboratory versus demonstration method on the pupils' final grade in general science.* Unpublished master's thesis, University of Washington, Seattle, Washington.

Shepardson, D. P., Moje, E. B., & Kennard-McClelland, A. M. (1994). The impact of a science demonstration on children's understandings of air pressure. *Journal of Research in Science Teaching, 31*(3), 243-258.

Shore, D. P. (1929). Demonstration laboratory versus individual laboratory in teaching high school physics. Unpublished master's thesis, George Peabody College for Teachers, Nashville, TN.

Shulman, L. S., & Tamir, P. (1973). Research on teaching in the natural sciences. In R. M. W. Travers (Ed.), *Second handbook of research on teaching: A project of the American Educational Research Association* (pp. 1098-1148). Chicago: Rand McNally & Company.

Silberman, R. G. (1983). A general chemistry demonstration: Student observations and explanations. *Journal of Chemical Education, 60*(11), 996-997.

Sorensen, L. L. (1966) Change in critical thinking between students in laboratory-centered and lecture-demonstration-centered patterns of instruction in high school biology. (Doctoral dissertation, University of Oregon, 1966). *Dissertation Abstracts International, 26*(11), 6567A.

Strehle, J. (1964). *The comparative achievement of seventh-grade exploratory science students taught by laboratory versus enriched lecture-demonstration methods of instruction.* Unpublished doctoral dissertation, University of Houston, Houston, TX.

Stuit, D. B., & Englehart, M. A. (1932). Critical summary of the research on the lecture-demonstration versus the individual laboratory method of teaching high school chemistry. *Science Education, 16*(5), 380-391.

Swafford, D. J. (1989). The effects of a science text and demonstration on conceptual change of high school students. *Dissertation Abstracts International, 50*(1), 3540A. (UMI No. 9003468).

Thompson, J., & Soyibo, K. (2002). Effects of lecture, teacher demonstrations, discussion and practical work on 10th graders' attitudes to chemistry and understanding electrolysis. *Research in Science & Technological Education, 20*(1), 25-37.

Trowbridge, L. W., Bybee, R. W., & Powell, J. C. (2000). *Teaching secondary school science: Strategies for developing scientific literacy.* Upper Saddle River, NJ: Merrill Prentice-Hall.

Van Horne, D. (1929). *An experimental comparison of demonstration and individual laboratory methods in high school chemistry.* Unpublished master's thesis, University of Southern California, Los Angeles, CA.

Waldman, A. S., Schechinger, L., & Nowick, J. S. (1996). A coordinated chemistry outreach program for thousands of high school students. *Journal of Chemical Education, 73*(8), 762-764.

Walter, C. H. (1926). *Can the demonstration method be made as effective as the laboratory method on the set up and happenings in the experiment.* Unpublished master's thesis, The University of Chicago, Chicago, IL.

Walter, C. H. (1930). The individual laboratory method of teaching physics when no printed directions are used. *School Science and Mathematics, 30,* 429-432.

Walter, C. H. (1930). The individual laboratory method of teaching physics when no printed directions are used. *School Science and Mathematics, 30,* 429-432.

White, J. R. (1943). A comparison of the group-laboratory and the lecture-demonstration methods in engineering instruction. Unpublished doctoral dissertation, School of Education, New York University, New York, NY.

White, R. T. (1994). Dimensions of content. In P. Fensham, R. Gunstone, & R. White (Eds.), *The content of science: A constructivist approach to it teaching and learning* (pp. 225-262). London: Falmer Press.

Wiley, W. H. (1918). An experimental study of methods in teaching high school chemistry. *Journal of Educational Psychology, 9,* 181-198.

Wilkinson, G. H. (1928). *A comparison of four methods of laboratory procedure in high school physics.* Unpublished master's thesis, University of Southern California, Los Angeles, CA.

Yager, R. E. (1966). Teacher effects upon the outcomes of science instruction. *Journal of Research in Science Teaching, 4*(4), 236-242.

Yager, R. E., Engen, H. B., & Snider, B. C. F. (1969). Effects of the laboratory and demonstration methods upon the outcomes of instruction in secondary biology. *Journal of Research in Science Teaching, 6*(1), 76-86.

www.ingramcontent.com/pod-product-compliance
Lightning Source LLC
Chambersburg PA
CBHW021239280526
45784CB00005B/2164